Protecting Our Environment

SUNY series in Global Environmental Policy

Uday Desai, editor

Protecting Our Environment

Lessons from the European Union

Janet R. Hunter
and
Zachary A. Smith

State University of New York Press

Published by
State University of New York Press, Albany

Printed in the United States of America

For information, address State University of New York Press,
194 Washington Avenue, Suite 305, Albany, NY 12210-2384

Production by Judith Block
Marketing by Michael Campochiaro

Library of Congress Cataloging-in-Publication Data

Hunter, Janet R., 1955–
Protecting our environment : lessons from the European Union /
Janet R. Hunter and Zachary A. Smith.
p. cm. — (SUNY series in global environmental policy.)
Includes bibliographical references and index.
ISBN 0–7914–6511–X (hardcover : alk. paper)—ISBN 0–7914–6512–8
(pbk. : alk. paper)
1. Sustainable development—European Union countries. 2. Environmental policy—European Union countries. I. Smith, Zachary A. (Zachary Alden), 1953– II. Title. III. Series.

HC240.9.E5H86 2005
333.7′094—dc22

2004029606

10 9 8 7 6 5 4 3 2 1

CONTENTS

Illustrations vii

Chapter 1: Introduction 1
Importance of the Study of the Environment for Humans 1
Overview of Europe 3
What/Who Is Europe 12
Introduction to Regime Theory 13
The Process of Regime Formation 20
Improving the Effectiveness of International Environmental Regimes 24
Applicability of Regime Theory to Environmental Problems 24
Conclusion 26

Chapter 2: The Organizational Structure of the European Union 27
Introduction 27
History of Integration of the European Nations 27
Treaties and Acts Amending the Founding Treaties 28
Governing Structure of the European Union 32
The European Environment Agency 49
The Role of the Public in EU Policy Making 52
Conclusion 55

Chapter 3: The European Environment 57
Introduction 57
History of Environmental Policy Development 58
Environmental Action Programs in the European Union 62
Summary 70

Chapter 4: Successes and Challenges in European Union Environmental Policy 71
Introduction 71
Case Studies 72
Implementation of Environmental Regulations 87
Monitoring Effectiveness 90
Summary 91

Chapter 5: A Comparative Evaluation of the European Union and the United States 93
Introduction 93
The United States Environment 93
Global Warming 99
Environmental Liability 104
Hazardous Waste 110
Solid Waste 111
Soil Erosion 113
Water Pollution 114
Land-Use Planning 116
Conclusion 117

Chapter 6: Conclusions 121
Introduction 121
Factors Leading to Environmental Policy Success in the European Union 121
Application to International Environmental Regimes 122
Conclusion 126

Appendix 1: Articles 249–256 of the Treaty on European Union: Provisions Governing the Institutions 127

Appendix 2: Articles 174–176 of the Treaty on European Union: Environment 133

Appendix 3: The Sixth Community Environment Action Program 136

Notes 167

Index 189

ILLUSTRATIONS

Figure 2.1	Text of Article 130R of Title XVI of the Treaty on European Union	31
Figure 2.2	Revisions Made to Article 130R of the Treaty on European Union by the Amsterdam Treaty	32
Table 2.1a	Representation of the Member States in the Governing Bodies of the European Union	35
Table 2.1b	Votes Within the Council of Ministers under Qualified Majority	36
Figure 2.3	Governing Structure of the European Union	36
Figure 2.4	Environment Directorate-General Mission Statement	37
Table 2.2	Number of Representatives in the Economic and Social Committee from each Member State	44
Table 2.3	Number of Representatives in the Committee of the Regions from Each Member State	46
Figure 2.5	Selected Excerpts from the Opinion of the Committee of the Regions on Amendments to Council Directives 85/337/EEC and 96/61/EC on Public Participation in the Development of Plans Relating to the Environment	47
Figure 2.6	Decision of the European Ombudsman on Complaints 271/2000(IJH)JMA and 277/2000(IJH)JMA against the European Commission	49
Figure 2.7	Decision of the European Ombudsman on Complaint 493/2000/ME against the European Commission	50
Figure 2.8	Selected Publications of the European Environment Agency	53

Figure 2.9 Minimum Standards for Consultation of Interested Parties by the Commission 55

Figure 3.1 Objectives of the European Commission on the Environment 58

Figure 3.2 Selected Environmental Conventions to Which the EU Is a Signatory 60

Figure 3.3 Selected Excerpts from the "Opinion of the Committee of the Regions on the Revision of the Treaty on European Union" Related to the Issue of Subsidiarity 63

Figure 3.4 Selected Excerpts from the Working Paper of the European Policy Center, "Beyond the Delimitation of Competences: Implementing Subsidiarity" 64

Figure 3.5 Fifth Environmental Action Program: Targets, Sectors, and Themes 67

Table 4.1 The European Environment Agency's Evaluation of EU Environmental Policies, 1998 72

Table 4.2 EU Programs for Integrating Environmental Concerns into the Energy Sector 81

Table 4.3 Energy Sources in Ireland, 1998 83

Table 5.1 The U.S. Environmental Protection Agency: 2000–2005 Strategic Plan 97

Figure 5.1 Two U.S. Superfund Sites 107

Figure 5.2 Two Incidents of Environmental Damage in the EU 109

Table 5.2 Principles Underlying the EU's Community Water Policy 116

CHAPTER 1

INTRODUCTION

IMPORTANCE OF THE STUDY OF THE ENVIRONMENT FOR HUMANS

The study of the environment is not just important for humans; it is essential. If we humans are to continue to survive and prosper, we must understand how our current use of environmental resources affects our ability to use and benefit from those resources both today and in the future. Our environment provides the sources of our livelihood and enjoyment of life, as well as sinks for the disposal of our waste products. Obviously, we humans could not survive in a vacuum, yet we often ignore the value that environmental resources hold, whether those be the foodstuffs that we consume, the lumber that we use to build our homes, or the psychic enjoyment that we receive from viewing a pristine valley.

The terms *environment* and *natural resources* are commonly used, but not always with clarity of meaning; therefore, we begin by defining these terms. We use 'natural resources' to refer to all land resources, that is, the soil and earth itself; the waters that run on the ground, under the ground, and in the sea; the air that encircles the earth, everything that grows on the land, such as crops, trees, wildlife; everything that grows in the seas, such as fish; and the resources that lie under the land, such as silver and iron. 'Environment' includes all natural resources but is more encompassing. Our environment is the capsule in which we live. It surrounds us, sustains and nurtures us, and provides us with not just the basic requirements of living but also with the amenities and pleasures associated with modern life. Thus, our environment includes our community, the structure and beauty of our buildings and city parks, the beaches and mountains in which we vacation, and the places in which we work and study. Our environment also includes our experiences with traffic jams, abandoned tenements, and landfills. Thus, 'environment' encompasses the entire context that defines how we live and how we perceive our lives. We cannot separate ourselves from the environment because we are a part of it; we provide input into the environment and take resources from the environment as part of our existence.

Natural resources and our environment determine how we survive as individuals and as communities. We build our homes on the land and rely on it to provide raw materials and groundwater. The earth provides the source of most human food, livestock feed, fiber, and fuel. Air is a resource on which all life on earth depends. Yet while our environment is so fundamentally necessary to our living, we also use it as a sink in which to place our wastes. Discovering how to strike a balance between using our natural resources and conserving them for the future is a lesson we are currently striving to learn.

Our environment is a significant determinant of human health. The environment is so important to our health that, worldwide, 25 to 30 percent of disease can be attributed to environmental factors.[1] In Europe, an estimated sixty thousand deaths per year may be due to long-term exposure to air pollutants; increased incidences of skin cancer have been linked to decreases in stratospheric ozone levels.[2]

The 1972 United Nations Conference on the Human Environment (also known as the Stockholm Conference) addressed the need for a set of common principles to guide the people of the world in preserving and maintaining their environment. Perhaps the Stockholm Declaration describes the relationship between man and nature best when it states,

> Man is both creature and moulder of his environment, which gives him physical sustenance and affords him the opportunity for intellectual, moral, social and spiritual growth. In the long and tortuous evolution of the human race on this planet a stage has been reached when, through the rapid acceleration of science and technology, man has acquired the power to transform his environment in countless ways and on an unprecedented scale. Both aspects of man's environment, the natural and the man-made, are essential to his well-being and to the enjoyment of basic human rights—even the right to life itself.[3]

Summary of the Book

This book addresses how the environment can be managed to sustain human life and pursuits, while protecting our environmental heritage for use today and in the future. We do this within the context of examining environmental policy in the European Union. This first chapter describes the European environment and the recent history of human life in Europe. We address the geological basis of the environment for human endeavor, as well as the forms of human activity in modern Europe. We then turn toward regime theory as an approach for considering methods to manage the environment. We discuss the value of regime theory for providing a basis for regulating and protecting our environment.

Chapter 2 examines the institutional environment of the European Union, that is the structures and rules that regulate environmental policymaking. Because the public guides institutional policymaking, we conclude chapter 2 with an examination of the role of the public in the European Union.

Chapter 3 focuses specifically on the European environment and actions designed to promote environmental protection. We examine the Environmental Action Programs established in the European Union (EU) to pursue environmental goals and the different foci of those programs over time.

Chapter 4 examines cases of challenges and successes in EU environmental policy. We consider cases involving fisheries, biodiversity and wildlife protection, transport, and economic development in rural areas. We provide an analysis of the cases, noting the particular economic incentives at play in each.

Chapter 5 provides a comparative analysis of the EU and the United States in environmental policy. We examine environmental liability policy, environmental protection, and water policy. We find that there are similarities but also differences between the two.

Chapter 6 provides our conclusions regarding the efforts of the EU to address environmental concerns. We return to regime theory as a method of addressing environmental protection in the international arena.

OVERVIEW OF EUROPE

The European Continent

The continent of Europe consists of forty independent countries and includes Russia west of the Ural Mountains and European Turkey. Of these European nations, fifteen are currently members of the European Union, although several eastern and central European nations could join the union over the next few years. In this section, we will discuss the European continent, considering both its physical and socioeconomic structure. We then review the major farming and manufacturing regions in Europe.

Physical Attributes. The European continent covers over 10 million square kilometers (4 million square miles), about 7 percent of the world's total land area.[4] Approximately three-fifths of Europe's land is close to sea level, at less than six hundred feet in elevation; one-third is between six hundred and three thousand feet in elevation. Because of its location, the continent receives the tempering effects of the marine influences from nearby oceans and seas. Continental Europe includes twenty-four thousand miles of coastlands; numerous islands lay off the coast.

The geography of Europe varies from mountains to plains, deserts, and valleys. The geography of northern Europe (Britain, Ireland, and Scandinavia) derives from ancient mountains. The large central part of Europe, from the Netherlands, Belgium and France to Germany, consists of lowlands. The dry southern portion of Europe consists of rugged mountains and plateaus. The eastern European countries consist of a relatively flat plateau. This area includes the former republics of the Soviet Union (Estonia, Latvia, Lithuania, Belarus, Ukraine, Moldova, Armenia, Azerbaijan, and Georgia). Eastern Europe also covers Bulgaria, Czech Republic, Slovakia, Hungary, Poland, Romania, and Albania. Europe has many rivers, most of which drain to the Atlantic; lakes make up only about 2 percent of the surface area.[5] Many rivers are interconnected by canals and are used in transport.

The climate of most of Europe is rather moderate. Western Europe has a maritime climate, with plentiful rainfall, (up to five hundred centimeters [two hundred inches] per year in Scotland and Wales), and mild temperatures. Central Europe has colder winters and warm summers; rainfall averages fifty to one hundred centimeters (twenty to forty inches).[6] Northeastern Europe has a dryer climate and long, cold winters and hot summers. The southern coast has a Mediterranean climate and moderate rainfall (twenty to forty inches). Winters in the south are mild and wet; summers are hot and dry.

Five types of vegetation are found in Europe. In the far north, including Scandinavia, Russia, and Iceland, where soils are swampy and poor, tundra is found. This area is covered by grasses and lichens and inhabited by reindeer, Arctic fox, and bear. The boreal zone of northern Russia and Scandinavia contains the most extensive forest in Europe, consisting of spruce, fir, pine, willow, and larch. Soils in this area tend to be acid and infertile. The deciduous zone from the British Isles to central European Russia consists of mixed forests of oak, maple, pine, beech, birch, elm, and linden. Soils in this area are suitable for crops, and agriculture is widespread. Grasslands naturally cover the steppes of Ukraine and southeastern Russia; however, the natural cover has been heavily replaced by croplands. Southern Europe's Mediterranean zone contains forests and scrub; however, most of the natural vegetation has been replaced by crops (notably, wheat, olives, and grapes).

Social Economic Environment

Europe's people represent approximately 15 percent of the total world population. The average birth and death rates of fourteen and nine per thousand, respectively, result in an annual population growth rate of 0.4 percent. However, many of the northern and western European countries have population growth rates near zero or below replacement rates.[7] Europe is the second most densely populated continent on the planet but has the lowest rate of natural increase. The average population density is approximately 65 per square kilometer (168 per square mile), although this varies from 50 per square kilometer (129 per square mile) in the Nordic countries (Norway, Sweden, and Finland) to 113 per square kilometer (396) in western Europe. The average for the United States is 25 per square kilometer (65) and for the world, 34 (88).[8] Most Europeans belong to the Caucasoid race. The population of Europe is highly urbanized, with the United Kingdom and Germany having the highest rates of urbanization and Albania and Portugal being the least urbanized.

There are approximately sixty languages native to Europe. The Germanic, Slavic, and Romance language groups are Indo-European, as are the Greek, Albanian, Celtic, and Baltic. The Finnish, Lapp, Karelian, Estonian, Hungarian, and Turkic languages are Ural-Altaic. Belgium has two language groups, the Walloons and the Flemish. Many Europeans are fluent in more than one language, with English, French, German, and Spanish being the most common second languages.

The standard of living in Europe is high. Literacy rates are also hig (over 95 percent) as education systems are well developed; the proportion of children in primary school is nearly 100 percent.[9] The average life expectancy ranges from 66.8 years for men in Poland to 80.7 years for women in Switzerland.[10] Calorie intake is the highest in the world, an average of thirty-five hundred calories a day per person.[11] The average per capita income in Europe is among the world's highest, although the southern European nations lag somewhat behind those of the north.

The religious affiliation of Europe's population varies, but the population is predominately Christian in religion. Catholicism is dominant in Spain, Portugal, Italy, France, Ireland, southern Germany, southern Belgium, and from Lithuania to Poland. Protestantism predominates in northern European countries of the United Kingdom, Scandinavia, northern Germany, and the Netherlands. The Northern states of Finland, Norway, and Sweden are predominately Lutheran. Groups of Muslims are found throughout Europe, primarily in southeastern European countries of Albania, Bulgaria, and European Turkey.[12] People who practice Judaism are scattered throughout Europe, but are most evident in the cities of central and eastern Europe.

Occupational activities vary, depending on the natural resources and industrial development of the area. Major occupations in the north are agriculture, fishing, forestry, mining, and industry. In the southern European countries of Spain, Italy, Greece, Portugal, and European Turkey, the major occupation is agriculture, but manufacturing is developing rapidly.

The governmental structure of the European states varies. Most of the European nations have socialist democratic governments. Great Britain is a constitutional monarchy, but sovereignty rests in Parliament. Norway, Sweden, and Denmark have figurehead monarchies.

Summary of Major Farming Regions and Types of Farming. Approximately one-third of Europe is considered arable. Another third is forested. Since the mid-1900s, about half the arable land has been used to grow cereals, principally wheat and barley. Less than one-fifth of the total land area of Europe is used for pasture. Major farm animals are pigs, sheep, and cattle; production varies across Europe. On the northern European plain pasture, oats, and barley predominate, along with milk and beef production, wheat, vegetables, potatoes, and flowers. Southern Europe produces fruits, vegetables, olives, and wines. Wheat and corn are major products in eastern Europe. Poland produces rye, potatoes, and livestock. Ukraine also produces livestock and potatoes, as well as sugar beets. Short growing seasons in the North limit agriculture to some extent; commercial forests are an important activity in this area. Forestry is concentrated in Russia and Scandinavia.

Agriculture employs less than one-eight of the workforce in most European countries. In the West and South, where soils are poor and rainfall is low, small family farms predominate. Agriculture in the South remains largely unmechanized compared to the countries of the North.

Fishing is an important occupation in the nations that border the seas. Based on total catch, Russia, Denmark, Norway, Iceland, and Spain are the leading fishing nations.[13] The most productive European fisheries are in the North Atlantic Ocean, the Norwegian Sea, the North Sea, and the Bay of Biscay.

Major Industries and Manufacturing Regions. Europe was the first region in the world to industrialize and develop a modern economy. The continent possessed iron ore for industry and waterpower and coal for energy. Approximately 35 to 40 percent of the labor force in most western European countries is engaged in industry and manufacturing activities; in southern and eastern Europe, agriculture is more important. The major European industrial powers are the United Kingdom, Germany, and France. The Netherlands, Belgium, Luxembourg, Switzerland, and Austria are also heavily industrialized. While manufacturing is still an important activity, from the early 1990s, the workforce has been turning more and more toward service industries such as banking and financial services. Europe represents half the world's international trade, with exports representing over 20 percent of the gross national product of the EU.

Western Europe manufactures high-quality machine and metal products, electrical goods, synthetic textiles, petrochemicals, motor vehicles, aircraft, computers, and consumer electronics. Southern Europe has limited resources of coal, iron ore, and petroleum. The states of eastern Europe have major mineral resources of coal, iron ore, petroleum, natural gas, and bauxite and produce basic industrial products, including iron, steel, and textiles. Among the eastern European states, the Czech Republic is the leading industrial nation.

Mineral industries in Europe are largely based on coal (mined in European Russia, the Ukraine, Germany, Poland, the Czech Republic, and the United Kingdom), although many coalfields are becoming exhausted. Iron ore is mined in Russia, the Ukraine, France, Sweden, and Spain. Petroleum production is confined to the North Sea area of the United Kingdom and Norway and to oilfields in Romania and Russia. The Netherlands, the United Kingdom, and Romania are the leading European producers of natural gas.

One-forth of the world's total electricity is generated in Europe, mostly from thermal-power plants; about 25 percent is from nuclear power plants. France is the highest consumer of nuclear energy, with about 70 percent of its electricity coming from nuclear production.[14]

Major European Environment Problems

Because of the long history of human population in Europe, as well as the denseness of population of some parts of Europe, a variety of environmental problems have become the focus of environmental policymaking in the EU. Next, we briefly describe some of the major environmental problems facing Europe.

Climate Change. Average annual air temperatures in Europe increased about 0.3 degrees centigrade during the 1990s, with climate change models predicting further increases of up to 2 degrees by 2100. This potential temperature increase brings with it the threat of droughts, more frequent and more intense storms, and changes in agricultural productivity. In order to avert such an increase in average temperatures, the emissions of greenhouse (carbon dioxide, methane, nitrous oxide, and halogenated compounds) by the industrialized nations will need to be reduced by at least 30 to 55 percent from 1990 levels by 2010.[15]

In Europe, the energy industry, through its burning of fossil fuels, was the biggest contributor of carbon dioxide emissions (the leading contributor to climate change) in the late 1900s and early 2000s, generating about one-third of total carbon dioxide emissions in the EU. Other emissions are produced in about equal proportions from household and commercial sectors, transport, and industry. In order to reduce the emissions of climate-changing emissions, the EU set the goal, under the 1997 Kyoto Protocol, of an 8 percent reduction by 2010. In addition, several member states (including Denmark, Austria, Finland, the Netherlands, and Sweden) have introduced energy/carbon taxes to reduce emissions.

Stratospheric Ozone Depletion. The depletion of the ozone layer in the late twentieth century resulted from enhanced levels of chlorine and bromine compounds in the stratosphere due to the use of chlorofluorocarbons (CFCs). CFCs are used as coolants in such modern devices as refrigerators, air conditioners, aerosol propellants, foaming and cleansing agents; ozone depleting substances are also present as halons used in fire extinguishers.[16] The Montreal Protocol and its extensions have been very successful in reducing the emissions of ozone-depleting substances into the atmosphere. However, because many ozone-depleting substances persist in the upper atmosphere, the ozone-depleting potential is not expected to reach its maximum until between 2000 and 2010.[17] After that, it will take many decades for the ozone level to recover.

Tropospheric Ozone and Summer Smog. Troposheric ozone refers to ozone concentrations in the troposphere, which exists from ground level to ten to fifteen kilometers (approximately six to nine miles) above the earth. Emissions from nitrogen oxides in industry and vehicles have led to an increase in the tropospheric ozone concentration of about three to four times since 1950.[18] High levels of tropospheric ozone negatively affect both human health and ecosystems. Frequently threshold levels (above which negative effects can be expected) are exceeded in many European countries, causing respiratory problems and reduced pulmonary function. High levels also damage vegetation, reducing yields and seed production.

The European Commission has developed several measures designed to reduce the level of ozone emissions in the troposphere. These include directives setting emission ceilings and reductions in emission levels; however, little success has been achieved as of 2003.

Acidification. Acidification refers to the effects of acid deposition from the emissions of sulphur dioxide, nitrogen oxides, and ammonia in the water and on soils. When fuels such as coal and oil undergo combustion in autos, industry, and power plants, emissions are carried into the air, where they remain for up to several days, and can be carried over long distances, affecting regions outside their area of origin. Acidification causes defoliation of trees, declines in fish stocks, and changes in soil chemistry. It also damages man-made structures, such as marble buildings and stained glass. Acidification above critical levels (the levels of deposition above which harmful long-term effects can be expected) remains a problem in about 10 percent of Europe, mostly in northern and central Europe.[19]

Success has been made in the reduction of sulphur dioxide emissions in the European Union, with a 50 percent reduction between 1980 and 1995. However, as the transport section is a major contributor of nitrogen oxide emissions, the growth in the use of automobiles has offset the benefits achieved from improving auto engines and exhaust systems. Thus, there continues to be a problem with acidification in the EU.

Chemicals. Huge numbers of chemicals are in use throughout Europe; the European Inventory of Existing Chemical Substances lists over one hundred thousand chemical compounds.[20] This is a significant concern because the chemical industry in western Europe grew faster than GDP in the late 1990s, yet the knowledge of how many chemicals are distributed and accumulate, as well as their impact on humans and the environment, have not been fully identified. In addition, the toxicity of many chemicals, especially when combined with other chemicals, is not fully known.

Heavy metals (notably, cadmium, mercury, and lead) and persistent organic pollutants (POPs), such as DDT, PCB, and dioxins, are of significant concern because these have been associated with reproductive disturbances in wildlife and humans, although causal links have not been established. Use of these chemicals has been addressed in the EU, and emissions of heavy metals are decreasing due to such steps forward as the reduction of lead in gasoline, cleaner technologies in the metal industry, and improved wastewater treatments. Control of the concentration of POPs in fishing waters has been less successful, partially due to the persistent nature of these chemicals.

Waste. Landfills are the dominant form of waste management in most European countries. About 420 kilograms (about 1,000 pounds) per person of municipal waste are generated each year in the EU member states, although the reduction of waste through recycling is growing. About 42 million tons per year of hazardous waste is generated every year, mostly in Germany and France. A large portion of the hazardous waste is the result of industrial activities, mining, and the clean up of contaminated sites. However, hazardous waste also results from consumer use of nickel-cadmium batteries, organic cleaning solvents, paints, and car engine oils.

Many legislative instruments have been implemented to harmonize waste legislation among the member states of the EU; however, the level of implementation, and therefore success, has varied among the states.

Continuing efforts emphasize cleaner technologies, improved product design, and material substitution.

Biodiversity. The pressure on the environment from human activity has been steadily increasing as human populations grow. Wild animal and bird species are declining under pressure from human encroachment and manipulation of the environment. Wetland loss has been significant in southern Europe due to land reclamation, pollution, drainage, and urbanization.

Urbanization and impacts from agriculture and forestry cause significant stress on the natural environment. While the total forested area is actually increasing, the use of exotic species is also increasing, replacing older, natural woodlands. More intensive agriculture, with higher use of chemical fertilizers and pesticides, has negatively affected the diversity of plant and animal life. The NATURA 2000 network (discussed in chapter 5) has been designed to protect diversity within the EU; however, implementation has been somewhat slow.

Inland and Marine Waters. Europe's waters are facing increasing pressure for various uses, from drinking and recreation to industry and agriculture. Groundwater quality faces its greatest threat from the runoff associated with nitrate use in agricultural production. Pesticide levels in groundwaters in some areas commonly exceed EU standards. In addition, pollution from heavy metals, hydrocarbons, and chlorinated hydrocarbons endanger water supplies. Several policy directives have addressed water pollution in the EU, including the recent Water Management Directive (discussed in chapter 4).

Marine waters face similar problems. Many waters have been overfished to the extent that some stocks are seriously depleted. (We discuss the North Sea fisheries in chapter 5.) In addition, eutrophication, that is, the deposition of mineral and organic nutrients in waters that promote the growth of plant life, especially algae, is a problem in many European seas. Eutrophication reduces the dissolved oxygen content of the water and leads to the extinction of organisms that naturally exist in the waters. The periodically occurring oil spills also harm marine life. Several EU initiatives have focused on sustainable development of coastal areas and the reduction of marine pollution.

Soil Degradation. In many areas, particularly around the Mediterranean, soil erosion and salinization are serious problems. Soil erosion has been addressed by policies aimed at reforestation. Salinization results from the use of water for irrigation, industrial, and urban development and causes lowered crop yields. Policies aimed at reducing salinization have not been developed in the EU as of 2003. The history of heavy industry in many areas has also led to soil degradation; over three hundred thousand contaminated soil sites have been identified. The EU's recent policies on environmental liability (discussed in chapter 4) have addressed the prevention of contaminated sites and the costs of cleaning up those sites.

Urban Environment. The urbanization of Europe's population has continued through the late twentieth century and early twenty-first century. As a result, many cities are experiencing lowered air quality, traffic congestion, and loss of green space. In addition, many European cities are straining their groundwater resources as the urban population grows.

Many European cities have focused on sustainable development, as addressed in Agenda 21; many have joined the European Sustainable Cities and Towns Campaign. Agenda 21, the Rio Declaration on Environment and Development, was adopted at the United Nations Conference on Environment and Development held in Rio de Janeiro in 1992. It provides a plan of action designed to address human impacts on the environment, noting, "In order to achieve sustainable development, environmental protection shall constitute an integral part of the development process and cannot be considered in isolation from it."[21]

The European Sustainable Cities and Towns Campaign began in 1994 following the First European Conference on Sustainable Cities and Towns, held in Aalborg, Denmark. The Aalborg Charter, developed at the Conference, provides a policy framework for the initiation of processes designed to lead to sustainable development plans at the local level. As of 2003, over 1860 local and regional European authorities are signatories to the Charter.[22] General funding for the campaign is provided by the Environment Directorate General of the European Commission, the Italian Association for Local Agenda 21, the City of Hannover, the City of Malmo, and the Barcelona City Council and Diputació. Local authorities, the members of the Steering Committee of the Campaign, and the European Commission provide for funding for specific projects.

Technological and Natural Hazards. Technological hazards include events such as accidents at nuclear installations, marine accidents, and industrial accidents. The European Council defines accidents as "sudden, unexpected, unplanned events, resulting from uncontrolled developments during an industrial activity, which actually or potentially cause serious immediate or delayed adverse affects to a number of people inside and/or outside the installation."[23] Data regarding technological hazards in Europe are not readily available, especially in the eastern and central European nations. However, these hazards are important because of their potential effects.

Events such as oil spills, while infrequent, can cause major environmental damage and result in huge clean-up costs. In order to respond to technological hazards, communications networks have been established and systems of emergency response developed in most European countries.

Natural hazards include events such as floods, blizzards, hailstorms, earthquakes, hurricanes, and heat waves. Damage from natural disasters has been increasing over the last fifty years, possibly due to human manipulation of the natural environment. Increased population densities and concentrations of industrial activity contribute to the potential impact of natural disasters in developed areas. While these hazards are obviously difficult to control, many European countries have taken steps to increase public awareness and knowledge of how to respond to such disasters.

Summary

As noted above, transport, energy, and agriculture are key forces that impact the environment. While EU environmental policies have addressed these issues, the development of the policies and their implementation has varied. These issues will be addressed as we progress through the book. However, before turning to these more specific issues, we first provide a general and brief overview of Europe.

WHAT/WHO IS EUROPE

Europe covers a variety of geological areas, as noted earlier. It includes numerous languages and cultures so that integration of these has at times been challenging. Nevertheless, the integration of European nations under the umbrella of the European Union has succeeded in achieving both economic and environmental goals. Later in the book we discuss the history of the integration of the European nations and the membership of the European Union today.

Members of the European Union

As of April 2003, there were fifteen member states in the European Union: Belgium, the Netherlands, Luxembourg, France, Italy, Germany, Britain, Ireland, Denmark, Greece, Spain, Portugal, Austria, Finland, and Sweden. Several nations of central and eastern Europe have petitioned to join the union. These candidate countries include Bulgaria, Cyprus, Czech Republic, Estonia, Hungary, Latvia, Lithuania, Malta, Poland, Romania, Slovakia, Slovenia, and Turkey.

Brief History of Region

Prior to the Congress of Vienna in 1815, which strove to settle various European political concerns, Europe was periodically disturbed by war generated by the various monarchs' quests for territory and power. In 1870 the German states joined to form a powerful union. By 1914, Germany had aligned with Austria-Hungary and was at war with France, Britain, and Russia. The outcome of the war led to the formation of several new nations in central and eastern Europe, as well as the creation of the Soviet Union. By 1939, another war was mounting in Europe, with Nazi Germany being defeated by an alliance of the Soviet Union, Britain, and the United States. The outcome led to the establishment of Soviet communism in many of the eastern and central European nations.

After World War II, some of the smaller European nations joined together to promote economic development. The first post–World War II integration of European nations began in 1948, when the Netherlands, Belgium, and Luxembourg created the Benelux Union as a customs union, providing for free trade among themselves and a common set of tariffs to the rest of the world. In 1951 the three Benelux states joined three others—France, Italy, and the Federal Republic of Germany (West Germany)—to sign the Treaty of Paris, forming the European Coal and Steel Community (ECSC). Integration continued when the Treaty of Rome, effective 1 January 1958, created two additional communities, the European Atomic Energy Community (or Euratom, designed to develop peaceful uses of atomic energy) and the European Economic Community (EEC, effective 1 January 1958). Besides eliminating customs duties among members and establishing a common customs tariff toward other nations, the EEC Treaty also provided for the creation of the European Investment Bank to facilitate economic expansion among the member nations. The ECSC, Euratom, and EEC agreed to be served by a single council of ministers, assembly, and court of justice. A few years later these three communities were merged into the European Communities, or as they are more commonly known, the European Community (effective 1 July 1967).[24] Additional members soon joined: in 1973, Britain, Ireland, and Denmark; in 1981, Greece; and in 1986, Spain and Portugal. Austria, Finland, and Sweden joined the new union in 1995. As of this writing plans are underway to add several central and eastern European countries, former members of the Soviet block, to the European Community (see above).

INTRODUCTION TO REGIME THEORY

As noted above, we will be examining environmental management in the European Union from the standpoint of regime theory. This section provides the theoretical basis for our study. It presents regime theory as an approach to evaluating the management of environmental resources. We first consider the nature of environmental resources. Then we examine theoretical approaches for managing environmental resources. Several approaches used for analyzing international regimes are considered. These include institutionalism, historical materialism, cognitivism, and neo-realism. The institutionalist approach is further subdivided into three alternatives: privatization, cooperative management, and public management. In the remainder of the book, we will develop our examination of the European Union as a regime.

The Nature of Environmental Resource Management

The need to collectively manage environmental resources used in common (common pool resources) has long been a challenge to mankind. Efforts to develop management systems for common pool resources date back hundreds of years. Records for the management of irrigation systems in Spain go as far back as 1435.[25] Thus, while these issues are not new, they are becoming more pressing as environmental degradation has worsened throughout the world in recent years.[26] Concern over environmental resources has resulted in numerous international agreements.[27]

There are different types or levels of common pool environmental resources. Many common pool resources are local in nature, such as the Spanish irrigation systems noted above. Some are more regional in scale, such as the Mediterranean Ocean, or Antarctica, while others are truly more global—the atmosphere and geostationary orbits, for example. Regardless of the level of environmental resource being considered, all share similar limits to their use that require some form of management.

Environmental resources may be either sources for production or consumption uses or sinks where wastes are deposited. Both source and sink uses can result in environmental problems. If too many people use a lake as a sink, eventually the sink will be unavailable for other uses, such as fishing. Thus it is necessary to develop a method for allocating resources for various purposes among competitive users. Underlying the various approaches to the management of our common environment lie the theoretical bases of analysis that attempt to explain how the various actors associated with the resource behave. Several theoretical approaches describing the behaviors associated with the management of environmental resources are examined in the next section.

Institutionalism

Institutionalism in political science emphasizes the role of institutional structures in imposing order.[28] Institutions, such as laws and rules, provide organization and avenues for control of political actors. In analyzing regimes associated with the environment, institutionalism emphasizes the structures associated with a regime, that is, norms, rules, and decision-making procedures, for determining the behavior of actors. Institutionalists focus on the interests or the context under which cooperation may be sought and attained. Actors are assumed to be utility maximizers, generally with incomplete information.[29]

There are three forms that institutional approaches for determining environmental resource use may take: privatization, cooperative (sometimes referred to as "collective") management, and public management. Public management of resources can occur at the national level, at the international level through an intergovernmental agency, or by a supranational unit. Each of these will be examined in turn.

Privatization. Privatization applies the institution of the market for determining the value, use, and allocation of environmental resources. Under a market-oriented approach to resource management, environmental resources would be privatized; that is, they would no longer be owned in common but would be owned by an individual or groups of individuals and would thus become private property. The owner(s) would then make any decisions regarding the use of the resource.

Privatization is realistically only applicable to local or regional common pool resources that can be effectively limited by boundaries, such as a lake or timber reserve. Resources that cannot be effectively contained and thus controlled would be subject to use by interlopers who do not own but wish to use the resource, resulting in a state similar to that of the uncontrolled commons.

Arguments supporting privatization hold that the assignment of property rights is the most efficient means of internalizing negative environmental externalities.[30] Privatization also provides incentives for the owner to appropriately manage the resource. However, simply changing the ownership characteristics of the resource will not necessarily improve resource use or prevent its destruction. The incentives associated with either private or public ownership must be evaluated to determine possible resource allocation issues that may occur.[31]

Cooperative Management. Privatization approaches may not be applicable to nonstationary resources (such as fisheries) or resources that are owned and used in common. In these cases, cooperative management by the users of common pool resources may be an effective approach to management of the resource. Those with firsthand knowledge of the resource are the most likely to be able to manage that resource appropriately in the long run as they have a stake in ensuring its availability for use in the future. Cooperative management may be able to effectively allocate use of the resource, while controlling encroachment on the resource.

Cooperative management has been shown to be effective in the management of small-scale common pool resources with clearly defined boundaries.[32] Examples include communal use of high mountain meadows of Switzerland, communal land use in Japanese villages, and community irrigation systems in the Philippines.[33] In these cases, the resource users have the potential to substantially harm each other, depending on the manner of their resource use but do not have the potential to harm others external to the community. In order for cooperative management to be viable, the users must be able to come to agreement regarding the use of the resource. Cooperative resource users must have common interests and believe that they can gain through cooperation.[34] In addition, cooperative users of the resource must be able to monitor resource use and be able to apply sanctions to violators.[35]

There are limits to group size for the cooperative management of resources.[36] In larger groups the independent actions of individuals tend to have lesser effects, so that larger groups are not as conducive to cooperative activities such as common pool resource management. In large groups it is easier to free-ride and otherwise violate the necessary terms of cooperative arrangements. There are also higher bargaining costs associated with voluntary arrangements involving larger numbers of members.[37]

Public Management of the Commons. As noted before, privatization and cooperative management of common pool resources may not be viable alternatives for the management of environmental resources. In cases where the boundaries of the resource are undefined (e.g., fisheries) or in cases where the results of resource use cross national boundaries (acid rain) intervention and/or management by a government or governments acting jointly, or even a supranational organization, may be necessary to achieve environmental management goals.

Most individual nations have environmental laws and regulations to deal with various environmental problems. The strength and degree of intervention associated with these varies from state to state and among various regions of the world. The developing nations tend to be much more tolerant regarding environmental quality, preferring to concentrate instead on development. Developed countries, on the other hand, have higher levels of wealth available to apply to environmental quality improvements. However, even within developed countries, various factions (e.g., nongovernmental organizations, business interests) generally have different objectives regarding environmental quality. When developed and developing countries come together to try to negotiate arrangements for the use and maintenance of environmental resources, the various perspectives and objectives they bring with them often lead to difficulties. The next section will consider joint management approaches at the international level from the institutionalist perspective of regime theory.

Supranational Management: International Regimes

This section will address the topic of international regimes. We will examine the nature of environmental regimes, how regimes can help solve environmental problems, and conditions that affect the effectiveness of regimes.

Regime theory is one approach that attempts to explain how actors behave. Regimes are social institutions that serve to direct the interactions of actors in defined situations. These behaviors may be based on norms, principles, rules, and/or decision-making procedures agreed upon by the relevant actors, in this case, nations. Generally international regimes cover specific issues areas and are determined by agreements among specific nations.[38]

International regimes, as suggested above, often arise in response to problems associated with environmental resources occurring at the international level.[39] Regimes offer a method for solving collective problems that cannot be solved by nations individually, but instead require some form of supranational structure. Regimes can provide avenues for improved environmental management by providing methods for collective choice, standard setting, rule making, compliance monitoring, and knowledge generation.[40] Regimes can also become institutions that shape their members' interests and lead to joint or collaborative projects; for example, members of the European Union often adjust national policies in response to Union directives.

The Role of International Organizations in Environmental Regimes

International organizations play a central role in international environmental regimes. For example, the United Nations has been particularly active in generating international responses to environmental concerns. International institutions provide contexts and sets of procedures through which representatives can develop common understandings and shared meanings.[41] International regimes may also stimulate societal learning by pooling existing capacities, by making explicit the goals of participating states, and by increasing awareness of environmental issues.[42]

Some scholars argue that international environmental concerns have resulted in the transference of power away from the nation-state as a locus of governance and toward local, regional, and particularly supranational levels of governance.[43] When addressing international or transboundary environmental issues, the relative importance of the individual nation-state decreases as the larger global issues take precedence. The World Commission on

Environment and Development report, *Our Common Future,* suggests that states cannot isolate themselves if they hope to effectively address environmental concerns.[44] Instead they must accustom themselves to increasing interdependence among nations.

The emergence of new international environmental problems and new coalitions of actors has also been found to influence policy formation at the domestic level by altering actors' interests or perceptions.[45] In China, for example, the influence of international technological information led to new understanding of the possibilities associated with responses to climate change and led China to take on a leadership position among developing countries.[46] Similarly in the negotiations on the Montreal Protocol, recognition of favorable international market conditions led commercial interests to support the protocol through efforts at both the domestic and international levels.[47] The CITES (Convention on International Trade in Endangered Species of Wild Fauna and Flora) regime influenced Zimbabwe's domestic policies. Zimbabwe's leaders realized that non-participation in CITES would be very costly (trade sanctions are imposed against violators). As a result, the African elephant was put on the domestic political agenda. However, once Zimbabwe became a member of the regime it tried to alter CITES goals.[48] (Weaker states often use their veto power to demand compensation or other forms of favorable treatment.)[49]

Other Theoretical Approaches

In addition to regime theory, there are several other approaches that attempt to explain how nation-state actors behave, for example, realism, neo-realism, and historical materialism. However, these are not as prevalent in environmental literature as regime theory, which we use as our foundation for analysis.

Creating Effective Regimes

The global environment can be considered an international commons, a resource over which two or more members of international society have an interest, but over which none has absolute jurisdiction, for example, the global climate system, open sea fisheries, or international waterways. There are three main options for dealing with the international commons.[50] The area may be enclosed, preventing overlapping jurisdictions. Alternatively, a supranational or world government may be created to oversee the use of these resources. Finally, codes of conduct can be introduced to direct appropriate behaviors on the part of nations sharing or competing for the use of the resource in question. Regimes function to provide these standards of behavior. This section describes theoretical approaches to creating and managing regimes. Although regimes can be evaluated on grounds of efficiency, equity, or ecological sustainability, the effectiveness of regimes is often viewed as of particular importance if the regime is to be successful.

Why Are Regimes Formed?

Regimes are formed in order to accomplish tasks that cannot be achieved by actors working individually. This becomes particularly relevant when dealing with the international commons—in situations where environmental resources are entirely or largely outside the jurisdiction of any individual state but valued by two or more of them as resources.[51] These actors realize that they must cooperate in order to achieve their goals. However cooperation is often possible only when parties with competing interests have an opportunity to generate options for mutual gain.[52] Therefore, regimes must create the necessary conditions for cooperation. For this reason, regimes may be more appropriate and useful when the relevant nations have similar goals and objectives for the environment.

Regimes may act in four capacities.[53] They may be predominantly regulative, focusing on the formulation of rules or behavioral prescriptions. They may act as institutions that primarily provide a locus for developing procedures for arriving at collective choices. They may be generative, developing new ways of thinking about problems. Finally, regimes can also become institutions that shape their members' interests and lead to joint or collaborative projects. For example, in environmental matters, members of the European Union have often adjusted national policies in response to union concerns.

Two types of explanation, structural and utilitarian, dominate the literature on regime creation. The structural view considers the potential for cooperation to be related to the structure of the international system. This view is that espoused by the neorealist perspective discussed above and often ties the possibilities for cooperation in the international arena to the existence of a hegemon. The hegemon's norms, principles, and rules are perceived to organize behavior in the system. The hegemon is expected to use its economic or military leverage over other states to bring them into regimes and to coerce them to comply with the constraints of the regime. The role of the United States in setting up trade and monetary regimes after World War II is often cited as an example of this type of regime formation.[54] However, this theory of regime formation sometimes fails to explain regime structure. For example, the regime for radio spectrum, organized when Britain was the hegemon, ran counter to the British preferences.[55]

The utilitarian explanation for regime creation is based on neoclassical economics and stresses cooperation on the basis of self-interest. Under this approach, regime members recognize that they are mutually vulnerable and that cooperation will improve joint welfare. Regimes serve to increase information and provide a set of stable expectations.[56] Utilitarian theory views the problem of the commons as a type of market failure. Consequently, the solution to an international environmental problem is to create a regime that can produce appropriate responses to market failure. Regimes function as institutions that minimize transactions costs and provide for monitoring and enforcement of regime provisions.[57] Solutions under this approach include such practices as offering transferable pollution rights, with the market determining a market price for such rights.[58]

THE PROCESS OF REGIME FORMATION

While the concerns of the relevant nations are fundamental to the regime formation process, other actors may also be involved. These include international and domestic nongovernmental organizations, which may influence the global environmental policy agenda by lobbying their own or other governments or by lobbying international negotiations. Corporations may also be involved, often lobbying their own domestic government or lobbying delegations to the negotiating conference. Generally, corporations have taken the position of weakening global environmental regimes, for example, in the whaling, ozone protection, logging in tropical forests, and toxic waste trade regimes.[59] In order for the reader to better understand how regimes develop and the potential roles of various actors, the stages in the regime formation process will now be described.

The first stage of the regime formation process is the evolution of the agenda. This stage focuses on the means by which issues come to international attention. This may occur through the efforts of nongovernmental environmental organizations or other citizen-based groups, as a result of scientific reports, or as a result of reports by various government agencies such as the European Environment Agency (discussed in chapter 3). In the case of environmental concerns, probable causes and the type of action required to correct the problem are identified in this stage. The second stage of regime formation is fact-finding. In this stage, the information needed for determining the scope and direction of the regime is gathered. Next, the bargaining process that creates the regime takes place. This negotiation stage can be critical and illustrates the necessity of developing an arrangement to which the participating nations will agree. Finally, additional bargaining and negotiations that strengthen the regime and reflect new knowledge of the environmental problem may take place. These may include the negotiation of protocols that establish concrete targets and timetables, amendments to existing agreements, or the adoption of stronger actions.[60] The final phase of regime formation is to operationalize the regime, completing the establishment of the regime.

However, the culmination of negotiations is only the beginning of the environmental regime thereby created. Once in place, efforts must be made to ensure that the procedures and rules created in the negotiations are carried out during the implementation phase. Some negotiations result in formal, binding agreements, whereas others provide for informal commitments. Some researchers have suggested that the more effective agreement may be the binding form because the signatories to the agreement are bound to meet the stipulations of the agreement. However, other research questions this assessment. The results of a number of case studies suggest that nonbinding agreements may actually be more effective in protecting the environment.[61] Nonbinding instruments may allow for more effective cooperation because states may be more willing to adopt this form of commitment when uncertainty regarding their ability to implement agreed-upon provisions exists. These studies also suggest that nonbinding agreements can be very effective when there is a small group of counties with similar concerns, allowing for deep integration, that is, active cooperation designed to achieve certain goals.[62] Binding agreements are often found to result in higher compliance rates, but this may simply be due to their less rigorous requirements. Countries may agree to binding treaties only when they are sure that they can meet the requirements of the agreement. Treaties and other legal structures on which regimes can be based are described below.

Legal Structures on Which Regimes Can Be Based

The legal structures utilized to govern regimes are often based on treaty.[63] Treaties establish clear rules and procedures for dealing with specific situations. Other, less binding legal structures include conventions, protocols, and codes of conduct. This section describes treaties and the other legal structures on which regimes can be based.

The Vienna Convention on the Law of Treaties governs the construction of global treaties. The convention provides that an agreement will take effect, or enter into force, when a sufficient number of parties have agreed to be bound by it. Various sections are included in a typical treaty. These include the articles that define the geographic scope of the treaty as well as key terms used in the document. Other articles define how the treaty will enter into force and how long it will remain open for signature. Articles may also call on parties to take "all appropriate measures" to address the problem, cooperate, and/or carry out certain provisions contained in the treaty.[64] Typically a treaty will include the rules of the regime regarding standards of behavior, the rules and procedures for allocating the resource in question, enforcement and compliance mechanisms, and arrangements for regime adjustments based on new knowledge regarding the resource. Generally arrangements for periodic meetings or conferences are also established. In addition, a secretariat, who is responsible for calling and supervising meetings, transmitting information, and ensuring coordination with other international organizations, is also provided for by the treaty. The treaty as a form of international agreement is the most binding.

In addition to treaties, other types of instruments may also be employed in regime development. These include conventions, framework conventions, protocols, and nonbinding codes of conduct. Conventions contain general agreements about basic principles or procedures. They may contain binding obligations or may be followed later by a more detailed legal instrument. Framework conventions are those that are negotiated in anticipation of later, more extensive text. Framework conventions establish a set of principles, goals, and formal mechanisms for cooperation on the issue. Protocols are preliminary memoranda that serve as the basis for a final convention or treaty. Typically they spell out specific obligations for the parties. Finally, nonbinding codes of conduct or guidelines (also referred to as "soft law") have been used in areas such as international pesticide and hazardous waste trade.[65] As noted above, these various legal structures may or may not necessarily lead to effective regimes. The next section provides a discussion of approaches to evaluating the effectiveness of regimes.

Evaluating the Effectiveness of International Environmental Regimes

Regime effectiveness can be considered from several perspectives. Commonly effectiveness is measured by the level of success of problem-solving efforts, that is, the extent to which the work of the organization attains its objectives. Some international problems may be solved more effectively than others because either the problem itself is simpler or because the problem-solving capacity of regimes differs. The problem-solving capacity of a regime may depend on the institutional setting, including the rules of the game, the distribution of power among the actors involved, and the skill and energy invested in designing and marketing cooperative solutions.[66]

Regime effectiveness may also be measured by the level of compliance among regime members. In this case, effectiveness is a function of the ability of the regime to influence its members to conform to the principles and implement the directives of the regime and its ability to monitor compliance with those principles. Compliance refers to whether countries adhere to the provisions of the accord as well as to the implementation measures; implementation refers to the methods that states take to make international accords part of their domestic law.[67]

Alternatively, effectiveness may be measured by the level of impact the regime has on the environmental resource in question. Under this approach to evaluating regime effectiveness, the emphasis is on determining whether the regime has, in fact, been successful in leading to environmental improvement. It is possible that a regime can be effective in generating compliance but lack the strength and force necessary to elicit the changes required to lead to progress in protecting the environment.[68]

Several characteristic regime features can influence the effectiveness of a regime. These include the process by which the regime is created and maintained, the scope and strength of the regime, the level of compliance among regime members, and the process of institutional learning through which collective behavior is modified in response to new learning or understanding.[69] Other factors influencing the effectiveness of a regime include the distribution of power among the actors involved and the skill and energy invested in designing and promoting cooperative solutions. Several factors that can limit the potential success of regimes are discussed in the next section.

Impediments to Forming Effective International Environmental Regimes. There are several factors that can impede the formation of effective international environmental regimes. The first is the implication (as espoused in the Brundtland Report) that solutions can be reached within the context of the current pattern of economic development.[70] The developing countries of the south view the current global environmental problems as the result of the actions of the dominant industrialized northern countries, and consequently, they expect the North to accept the responsibility for environmental damage. The southern countries may perceive that there is little or nothing to gain by signing on to international environmental agreements.

Other problems may also limit the potential success of international environmental regimes. For instance, some international treaties establish representation and voting procedures that do not guarantee all countries and interests equitable treatment. Similarly, an imbalance of the influence of science and politics in negotiations may not lead to effective agreements. In addition, linkages between environmental issues and other policy issues that could improve the effectiveness of environmental agreements are generally not developed.[71] These various impediments tend to lead to least-common-denominator results, rather than optimal treaty agreements.

A potentially significant impediment to the formation of effective international environmental agreements is the possibility that there are conflicting interests between the individual state's desires and the requirements necessary to create effective global regimes.[72] Differing interests can lead to internal and external pressures on national negotiating committees and make it difficult for states to reach an agreement. In addition, scientific and political considerations are often unbalanced, and countries may use technical or scientific information inappropriately to advance their short-term goals. Given the lack of scientific certainty regarding environmental resources, countries may lack sufficient scientific knowledge to make good decisions. Thus several potential pitfalls can diminish possibility of creating successful environmental regimes. However, there are several measures that can be used to help to achieve effective environmental regimes. These are discussed below.

IMPROVING THE EFFECTIVENESS OF INTERNATIONAL ENVIRONMENTAL REGIMES

Several measures can help strengthen the effectiveness of international environmental regimes. These include convincing national leaders to support efforts to protect and improve the environment, strengthening the national agencies responsible for supervising the implementation of environmental regulations, supporting international and national nongovernmental environmental organizations, and building epistemic communities.[73] An evaluation of the record regarding compliance with international environmental agreements indicates that more emphasis should be placed on improving information and reporting systems and inducing compliance through negotiation and incentives, rather than through the threat of punishment.[74]

As an alternative to the traditional approach to creating international agreements, a multistep process has been proposed for improving the success of international agreements. This alternative approach is based on work done as part of the Salzburg Initiative, sessions that bring together international and national leaders to discuss and debate the merits of possible reforms to traditional approaches to the creation of global environmental treaties. Recommendations developed by the institute include providing prenegotiation assistance to individual countries to develop informed perspectives, expanding the roles for nongovernmental interests, and encouraging the media to play a more educative role.[75]

APPLICABILITY OF REGIME THEORY TO ENVIRONMENTAL PROBLEMS

As we noted earlier in this chapter, human activities depend on the natural gifts of our planet. If our activities are to be sustained in the future, we must ensure that the natural systems on which our human systems depend are protected and preserved. The carrying capacity of our environment is limited; it cannot support unlimited amounts of human activity. Therefore, we need to discover ways to engage in needed human activities while still maintaining the environment on which we depend for our livelihoods.

Although most individual nations have environmental laws and regulations to deal with various environmental problems, in many cases the boundaries of the resources may cross national boundaries, making international management necessary for successful management of the environmental resource. Many international environmental regimes have attempted to protect and maintain various sectors of our environment. As we noted above, these have included such diverse areas as deep seabed minerals, marine pollution, stratospheric ozone, and geostationary orbits. Some of these have been more successful than others; many times political and economic decisions take priority over environmental concerns. If we are to maintain and protect our natural resources and environment for the future, then we must discover how to develop more effective international environmental regimes. We cannot simply continue as we have been; we must become more successful in managing our environment.

The European Union as a Regime

The European Union is a regime as defined above. It is a group of nations that have voluntarily come together to achieve specific goals by developing sets of principles, rules, and decision-making procedures. The EU has achieved considerable success in meeting many of its goals; thus it is an appropriate case for analysis of factors that contribute to successful regimes.

We have chosen to examine the European Union as a regime because by understanding the environmental challenges faced by the EU and the approaches taken to successfully meet those challenges, we will better understand the structures and incentives that will allow other international environmental regimes to be successful. As we discussed above, this is particularly important when examining international environmental regimes. While the EU did not begin as an environmental regime, it has altered its policies to make environmental considerations a part of all policy-making decisions.

An examination of the European Union will aid us in understanding the requirements for developing and maintaining successful international environmental regimes. Our understanding of the significant factors that result in successful regime policies and the effectiveness of those policies is vital if we are to be able to protect and maintain our environmental resources for the future.

As we will see in chapter 6, the EU still struggles to implement and enforce environmental regulations. However, as we will also see, there has been great success in many environmental areas. The EU has been able to integrate environmental concerns into the overall policy-making structure, focusing on the need to protect and preserve natural resources. The role of the public, as we shall see, has been important in focusing the orientation of the government toward the environment.

CONCLUSION

The nature of many environmental resources results in a common pool problem. When the commons is on an international scale, regimes have often been created as devices to handle the problem of the common pool resource use that nations cannot manage individually. This chapter has reviewed theoretical approaches to regimes and factors affecting the effectiveness of regimes. As this chapter suggests, the formula for creating successful regimes is still under review. Many factors influence the success of a regime, including the actors involved, the stakes of the game, and the particular environmental problem at issue. In chapter 2 we will examine the European Union as a regime from an institutional perspective. We will review the history of the EU, along with the various treaties that have led to the current EU. A review of the policy-making structure, focusing on environmental policy-making, will also be provided to illustrate the nature of the institutional setting of the EU.

CHAPTER 2

THE ORGANIZATIONAL STRUCTURE OF THE EUROPEAN UNION

INTRODUCTION

This chapter will describe the organization of the European Union. As we noted in our discussion of regime theory in chapter 1, the institutional structure of a regime is important because the rules, norms, and decision-making procedures of the regime help determine the behavior of actors within that regime.

The European Union is composed of independent nations who maintain political and legislative power at the national level; thus, we begin with a history of the creation of the EU. We then continue with a discussion of the organization of the governing structure and the legislative process in the EU. We focus on the laws and treaties that govern policymaking in the EU, that is, the legal structures governing the regime, particularly as these apply to environmental policy. Good policy making requires good information. The European Environment Agency (EEA) is the EU's environmental data-gathering and disseminating agency; thus, we conclude this chapter with a description of the EEA.

HISTORY OF INTEGRATION OF THE EUROPEAN NATIONS

The first post–World War II integration of European nations began in 1948, when the Netherlands, Belgium, and Luxembourg created the Benelux Union as a customs union, providing for free trade among themselves and a common set of tariffs to the rest of the world. In 1951 the three Benelux states joined three others, France, Italy, and the Federal Republic of Germany (West Germany), to sign the Treaty of Paris, forming the European Coal and Steel Community (ECSC). Integration continued when the Treaty of Rome, effective January 1, 1958, created two additional communities, the European Atomic Energy Community (or Euratom, designed to develop peaceful uses of atomic energy) and the European Economic Community

(EEC, effective 1 January, 1958). Besides eliminating customs duties among members and establishing a common customs tariff toward other nations, the EEC Treaty also provided for the creation of the European Investment Bank to facilitate economic expansion among the member nations. The ECSC, Euratom, and EEC agreed to be served by a single council of ministers, assembly, and court of justice. A few years later these three communities were merged into the European Communities, or as they are more commonly known, the European Community (effective 1 July, 1967).[1] Additional members soon joined: in 1973, Britain, Ireland, and Denmark; in 1981, Greece; and in 1986, Spain and Portugal. Austria, Finland, and Sweden joined the new Union in 1995. As of this writing plans are under way to add several central and eastern European countries, former members of the Soviet block, to the Community.

The Treaty of Nice

Although not of direct importance to the development of European environmental policy, the Treaty of Nice is fundamentally important in the history of the development of the EU and hence is included here.

The Treaty of Nice, signed in February 2001, is designed to provide for the enlargement of the EU by the addition of countries of central and eastern Europe.[2] It provides measures for increasing representation in the various legislative organizations when new member states join the union. In addition, the treaty provides for enhanced cooperation among the member states by adding provisions that allow a minimum of eight member states to cooperate in certain fields, such as, criminal investigations, with other members free to join later if they desire.[3]

The treaties discussed above outline the historical development of the EU as an institution. These treaties—along with others described below—provide the basis for the governing structure of the Union.

TREATIES AND ACTS AMENDING THE FOUNDING TREATIES

Since 1980, several treaties and acts have amended the founding treaties in order to deal with issues that have arisen since the time of the creation of the European Communities in the 1940s and 1950s.[4] These issues include such concerns as the environment, the integration of the economies of the communities of the EU, and citizen rights in the Union. These treaties and acts, particularly as they relate to environmental issues, are discussed below.

The Single European Act

The Single European Act (SEA) was the result of a number of additions and amendments to existing treaties between the European States. The

SEA laid out a plan to integrate the economies of the EC and identified several goals, including the elimination of internal, physical, technical, and fiscal borders. The SEA addresses cooperation in monetary matters but does not specifically address the creation of a common currency. The integration of the economies of the European Community was planned to create prosperity and a new competitiveness on world markets. The SEA was adopted in 1983 and expanded in 1984 and 1985.

The SEA is the first major multilateral European treaty to specifically consider the environment as a community issue. Title VII of the SEA defines community principles of action on the environment and sets the goal of improving and preserving the environment and public health and using natural resources prudently. Articles 130r, 130s, and 130t of the SEA contain the basic principles of community environmental policy. These include protecting the quality of the environment and the health of people, using natural resources rationally, and promoting, at the international level, measures to deal with environmental problems. According to the SEA, the environmental principles that form the basis for policy making include the precautionary principle and preventive action, the polluter pays principle, and the rectification of damage at the source. The precautionary principle is to be used in situations "where scientific data are insufficient, inconclusive or uncertain and where . . . the possible effects on the environment or human, animal or plant health may be potentially dangerous and inconsistent with the chosen level of protection."[5] Basically, the precautionary principle argues that if there is evidence of harm, even though there may be some doubt about causality, the prudent thing to do is to act. The polluter pays principle simply argues that the costs of pollution cleanup and prevention should be paid by the polluter. In addition, the SEA states that environmental principles are to be included in the definitions and implementation of policy in other areas that might be addressed by the EC. Finally, the treaty also allows any member state to implement more stringent policies than the community, if it desires.

The Treaty on European Union

The second treaty we will examine here that has played an important part in the development of European environmental policy is the Treaty on European Union (also known as the Maastricht Treaty). The Treaty on European Union, ratified in 1992, is most important in European history for providing a plan to incorporate the European Communities into a European Union. Amendments related to the environment were added, however, making environmental policy, including action at both global and regional levels, a basic EU policy objective. Although the Treaty on European Union theoretically joined the member states to a higher level of cooperation and economic convergence, the member state remains the fundamental institution for much environmental policy innovation and all environmental policy implementation.[6]

The Treaty on European Union reinforced the environmental efforts begun in the Single European Act. Of particular significance to environmental concerns is the goal of "sustainability." Sustainable development is incorporated as a fundamental policy objective, by calling for, "economic and social progress which is balanced and sustainable."[7] Perhaps most important, and some would argue now most ignored, Article 130s of the treaty requires that environmental protection be integrated into other policies. (See Figure 2.1 for the text of Article 130R.)

This approach resulted in the use of elements designed to heighten awareness of environmental impacts and to force polluters to internalize the costs of their pollution (i.e., to force polluters to pay the costs resulting from their polluting activities).

The Treaty on European Union recognizes that environmental concerns extend beyond state boundaries and that any policy measures must envelop this reality. Section 2 of Article 130d specifically states, "Environmental protection requirements must be integrated into the definitions and requirements of other Community policies."[9] Article 130r takes another step in this direction, promoting measures at the international level.[10] The community recognizes that international cooperation is a necessary step toward solving with environmental problems that surpass regional boundaries.

Although the Treaty on European Union supports environmental protection measures, it may also, at times, hamper the development and application of environmental legislation. In particular, the subsidiarity principle has been noted as causing problems for environmental policymaking. Article 3b of the Treaty on European Union describes the principle of subsidiarity thusly:

> In areas which do not fall within its exclusive competence, the Community shall take action, in accordance with the principle of subsidiarity, only in and insofar as the objectives of the proposed action cannot be sufficiently achieved by the Member States and can therefore, by reason of the scale or effects of the proposed action, be better achieved by the Community.[11]

The application and use of the subsidiarity principle leads to differences of opinion between the EU and member states about the appropriate level for environmental policy making. The issue of subsidiarity is discussed in further detail in chapter 3.

The Treaty of Amsterdam

The Treaty of Amsterdam builds on and modifies the Treaty on European Union. Its goals are to provide for a more effective and more democratic

Figure 2.1.
Text of Article 130R of Title XVI of the Treaty on European Union

TITLE XVI
ENVIRONMENT
ARTICLE 130R

Community policy on the environment shall contribute to pursuit of the following objectives:
Preserving, protecting and improving the quality of the environment;
protecting human health;
prudent and rational utilization of natural resources;
promoting measures at international level to deal with regional or worldwide environmental problems.

Community policy on the environment shall aim at a high level of protection taking into account the diversity of situations in the various regions of the Community. It shall be based on the precautionary principle ad on the principles that preventive action should be taken, that environmental damage should as a priority be rectified at source and that the polluter should pay. Environmental protection requirements must be integrated into the Community's other policies.

In preparing its policy on the environment, the Community shall take account of:
- ➢ Available scientific and technical data;
- ➢ Environmental conditions in the various regions of the Community;
- ➢ The potential benefits and costs of action or lack of action;
- ➢ The economic and social development of the Community as a whole and the balanced development of its regions.

With their respective spheres of competence, the Community and the Member States shall cooperate with third countries and with the competent international organizations. The arrangements for Community cooperation may be the subject of agreements between the Community and the third parties concerned, which shall be negotiated and concluded in accordance with Article 228.[9]
The previous subparagraph shall be without prejudice to Member States' competence to negotiate in international bodies and to conclude international agreements.

Source: *Consolidated Version of the Treaty Establishing the European Community*, Official Journal C340.10.11.1997, 173–308. Accessed 27 March 2002. Available from http://www.europa.eu.int/eur-lex/en/treaties/dat/ec_cons_treaty_en.pdf

union by giving citizen's rights more weight in policy, by improving freedom of movement, increasing the efficiency of union institutions, and increasing Europe's role in world affairs.[12] The treaty strengthens that commitment of the EU toward sustainability by inserting the concept of sustainable development into the treaty within the preamble and in the objectives of the EU Treaty, as well as in Article 2 of the treaty.[13] Specifically, it provides that the

EU shall include among its objectives "to achieve balanced and sustainable development" (Article 2).[14] (Significantly, no concrete definition of "sustainable development" is provided.) The Amsterdam Treaty also revised Article 130, sections 2 and 4, of the Treaty of European Union to provide for actions to be taken at the member state level to meet environmental protection requirements. The changes made by the Amsterdam Treaty to the Treaty on European Union, Article 130 are shown in Figure 2.2.

Figure 2.2.
Revisions made to Article 130R of the Treaty on European Union by the Amsterdam Treaty

Treaty on European Union
Article 130r

1. (Paragraph unchanged.)
2. Community policy on the environment shall aim at a high level of protection taking into account the diversity of situations in the various regions of the Community. It shall be based on the precautionary principle and on the principles that preventive action should be taken, that environmental damage should as a priority be rectified at source and that the polluter should pay. Environmental protection requirements must be integrated into the Community's other policies.
3. (Paragraph unchanged.)
   One revision: Article 228 (in Treaty on European Union) changed to corresponding article in Amsterdam treaty, Article 300.

Amsterdam Treaty
Article 174

2. Community policy on the environment shall aim at a high level of protection taking into account the diversity of situations in the various regions of the Community. It shall be based on the precautionary principle and on the principles that preventive action should be taken, that environmental damage should as a priority be rectified at source and the polluter should pay.
   In this context, harmonization measures answering environmental protection requirements shall include, where appropriate, a safeguard clause allowing Member States to take provisional measures, for non-economic environmental reasons, subject to a Community inspection procedures in accordance with Article 300.

Source: Euroconfidentiel, *The Rome, Maastricht and Amsterdam Treatie: Comparative Texts* (Belgium: Euroconfidential, 1999) p. 143–4.

GOVERNING STRUCTURE OF THE EUROPEAN UNION

As we have seen, the organizational structure and legislative process of the EU is based on several treaties. The most important of these are the Treaty of

Paris (1951), the Treaty of Rome (1957), the Single European Act (1986), the Treaty on European Union (1992), the Treaty of Amsterdam (1999), and the Treaty of Nice (2001).

Governing Bodies

The Treaty of Rome created four institutions for overseeing and implementing provisions. These institutions are the European Commission, the Council of Ministers, the European parliament (formerly the Assembly), and the Court of Justice. Representation on each of theses bodies is determined by several factors, including population. (Figure 2.3 provides the population of each country, the percentage of EU population each country represents, and the number of representatives from each country in the commission, Council of Ministers, and European parliament.) The Economic and Social Committee, the Committee of the Regions, the European Ombudsman, and the European Council assist these four major units in their work. (Figure 2.4 provides a schematic of the organization of the governing structure of the European Union.) The organization, membership, and purpose of each unit are described below.

The European Commission. The role of the European Commission is executive in nature and includes the initiation, supervision, and implementation of policies. The commission is charged by Article 155 of the Treaty of Rome to "ensure that the provisions of this treaty and the measures taken by institutions pursuant thereto are applied."[16] The commission consists of a College of Commissioners along with a civil service, which is composed of administrators, researchers, and translators. The commissioners are nominated by national governments, but according to the Treaty of Rome, "they shall neither seek nor take instructions from any government or from any other body" as they are to act in the interest of the EU as a whole.[17] France, Germany, Italy, Britain, and Spain each have two commissioners; the other states each have one. The commissioners are appointed for five-year renewable terms. The president of the commission is chosen by agreement among the EU Heads of State or Government and serves as president for one term, with the option for renewal.[18]

To initiate legislation, the commission submits proposals to the Council of Ministers (discussed below). These proposals can later be amended by the commission itself or by the council. Input at the proposal generation stage can be provided by interest groups, experts, and national representatives. The commission also drafts the annual budget, in conjunction with the European parliament, for discussion in the council and in Parliament. The commission may also refer cases to the European Court of

Justice. Most commission decisions are determined by simple majority vote. (EU treaties determine the voting procedure used in an area.)

There are thirty-six different Directorates-General (DGs) in the EU, who are responsible for initiating legislative proposals in their areas.[19] Each DG is headed by a director-general, a position equivalent to a top-level civil servant. Each director-general reports to a commissioner who is responsible for that DG. Some of the twenty commissioners, thus, have responsibility for more than one DG.

The Environment Directorate-General is responsible for environment and nuclear safety and is generally referred to by its number, DG-XI. The DG initiates proposals for legislation to be submitted to the commission. When developing legislation, the DG holds discussions with representatives of various segments of society (which may include government, environmental nongovernmental organizations, industry, and/or other special interest groups) in order to gather input from various competing interests. The Environment DG is also responsible for ensuring that EU environmental legislation is correctly employed by the member states and may initiate legal actions against those who do not comply. In addition, the Environment DG is the representative of the EU at the international level.

The office of DG-XI has a permanent staff of 150, who are assisted by an equivalent number of temporary staff and/or consultants. The office also houses several committees made up of experts who provide information and advice. Figure 2.4 provides the Environment Directorate-General's mission statement.

The Council of the European Union. The role of the Council of the European Union, more commonly known as the Council of Ministers, is to make decisions on proposals brought to it by the commission. The Council of Ministers is secondary to the commission and is the domain in which member states can introduce their concerns. The Council of Ministers is the only institution in the EU in which the national governments of the member states are represented. The council has fifteen members, one member from each state.

The Council of Ministers is unusual in that it does not have a permanent membership. Instead, representation depends on the subject being discussed. For example, the energy ministers will represent their states in matters of energy, the agricultural ministers in agricultural matters, and so on. The council meets once or twice a month in its various forms. The chairmanship of the council rotates every six months. Depending on the matter under discussion, council decisions may be based on either unanimity or qualified majority vote; treaties determine which form is to be used. Unanimity occurs when all member states agree (or all abstain). Matters

Table 2.1.a

Representation of the Member States in the Governing Bodies of the European Union

	Total Population of Member State	*Percentage of EU Population*	*Members of the Council of Ministers*	*Weight of Votes in Council of Ministers*	*Members of the Commission*	*Members of Parliament*
Germany	83,251,851	21.9%	1	10	2	99
France	59,765,983	15.8%	1	10	2	87
Italy	57,715,625	15.2%	1	10	2	87
United Kingdom	59,647,790	15.7%	1	10	2	87
Spain	40,037,995	10.6%	1	8	2	64
Belgium	10,274,595	2.7%	1	5	1	25
Greece	10,645,343	2.8%	1	5	1	25
Netherlands	16,067,754	4.2%	1	5	1	31
Portugal	10,066,253	2.7%	1	5	1	25
Austria	8,169,929	2.2%	1	4	1	21
Sweden	8,875,053	2.3%	1	4	1	22
Denmark	5,368,853	1.4%	1	3	1	16
Ireland	3,883,159	1.0%	1	3	1	15
Finland	5,183,545	1.4%	1	3	1	16
Luxembourg	448,569	0.1%	1	2	1	6
Total	379,402,297	100.0%	15	87	20	626

Table 2.1.b
Votes Within the Council of Ministers under Qualified Majority
(Qualified majority requires sixty-two votes out of the total of eighty-seven to be approved.)

Country	*Weight of votes*
Germany, France, Italy, United Kingdom	10
Spain	8
Belgium, Greece, Netherlands, Portugal	5
Austria, Sweden	4
Denmark, Finland, Ireland	3
Luxembourg	2

Source: Adapted from "Countries of the World," Infoplease.com, <http://www.infoplease.-com/ipa/A0107895.html> 26 March 2002; Members of Council, Commission, and Parliament: "Representations of the European Commission in the Member States," <http://www.europa.eu.int/comm/offices.htm#dk> 26 March 2002.

Figure 2.3
Governing Structure of the European Union

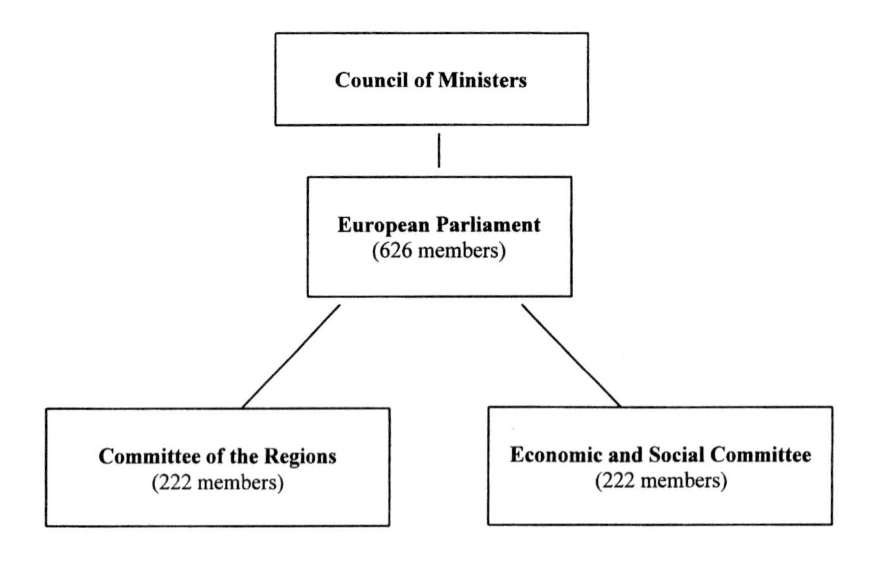

Figure 2.4.
Environment Directorate-General Mission Statement

To maintain and improve the quality of life through a high level of protection of our natural resources, effective risk assessment and management and the timely implementation of Community legislation.

To foster resource-efficiency in production, consumption, and waste-disposal measures.

To integrate environmental concerns into other EU policy areas.

To promote growth in the EU that takes account of the economic, social and environmental needs both of our citizens and of future generations.

To address the global challenges facing us notably combating climate change and the international conservation of biodiversity.

To ensure that all policies and measures in the above areas are based on a multi-sectoral approach, involve all stakeholders in the process and are communicated in an effective way."[20]

related to the environment, common foreign and security policy, immigration policy, taxation, and economic and social cohesion policy are usually decided by unanimous voting. When the council takes decisions by qualified majority, votes are weighted as follows: Germany, France, United Kingdom, and Italy, each weighted by ten; Spain, weighted by eight; Belgium, Greece, the Netherlands, and Portugal, weighted by five; Austria and Sweden, each weighted by four; Denmark, Finland, and Ireland, each weighted by three; and Luxembourg, weighted by two.[21] (See Table 2.1a and b.) Qualified majority requires at least sixty-two votes cast for a decision (from the possible total of eighty-seven votes resulting from weighting each country's votes).[22] Qualified majority voting is generally used for matters related to the internal market.

Legislative Instruments Used by the Commission and Council of Ministers

The European commission and/or the Council of Ministers can issue five different instruments to influence or change the law of member states. These are regulations, directives, decisions, recommendations, and opinions, the first two being the most common. All forms are provided for in Article 249 of the Treaty on European Union. Both regulations and directives are initiated by the commission and adopted by the Council of Ministers. Often an opinion is received from the European parliament; one is sometimes received from the Economic and Social Committee, the Committee of the Regions, or another community body.

Regulations are more general and apply in all member states. They establish direct rights and impose duties on private parties without interference of national law. Regulations generally can be adopted in the field of environmental protection only when the subject concerns external trade. Regulation 348/81, for example, applies to limitations on the importation of products derived from whales.

Directives are binding to the member states to which they are addressed but allow the member states to determine the method of implementation. Generally the member states must implement the directive in national law within two years. Almost all EU legislation involving environmental concerns has been in the form of directives. Decisions may be addressed to a government, an enterprise, or an individual. Decisions are binding only upon those entities toward whom the decision is addressed. Recommendations and opinions are simply that and are not binding.

Throughout the 1960s and 1970s environmental directives were focused on "command and control" or "standards and enforcement" regulatory strategies and were oriented toward setting limits on specific pollutants, such as motor vehicle exhaust. By the mid-1980s, the orientation had moved toward the use of broader strategies, such as the use of environmental taxes and per unit fees for the discharge of pollutants. This reflects a subtle shift in regulatory policy that was (and is) occurring throughout the more developed world, notably in the United States.

The instruments discussed above are initiated at the proposal level through four different procedures: consultation, cooperation, codecision, and assent. The two most commonly used procedures are consultation and codecision. Under consultation, the commission submits a proposal to the Council of Ministers. The council then must obtain the opinion of the European parliament, the Economic and Social Committee, and the Committee of the Regions. The decision is made by the council and requires either a qualified majority or a unanimous vote. The decision rule depends on the field involved; for example, qualified majority is used in agriculture and unanimous voting for the field of taxation.[23]

Decision-making procedures were adjusted somewhat as a result of the Treaty on European Union. "Majority" was selected over "unanimity" as the "standard" procedure in most decision making. However, for Article 130s legislation majority decision making was maintained as the standard except for three cases, two of which are related to environmental concerns (the third to fiscal matters):

- town and country planning, land-use measures (except waste management),
- measures of a general nature and regarding water resources management, and
- measures significantly affecting a Member State's choice between different energy sources and the general structure of its energy supply.[24]

The codecision procedure was brought into use with the Treaty on European Union and provides additional power for the European parliament. This procedure is used for decisions regarding the internal market, transport policy, environmental policy or research programs.[25] Decisions are adopted under the codecision procedure if both the European parliament and the council agree by qualified majority. If the council and European parliament do not agree regarding proposals, a Conciliation Committee, which consists of representatives from both groups, is formed to negotiate a mutually acceptable compromise.

Under cooperation, the European parliament is given additional rights in legislation, with a few exceptions. These additional rights include being consulted and having the opportunity to propose amendments. For Parliament to propose amendments, an absolute majority vote is required.

The assent procedure primarily governs agreements with nonmember states but has been extended to include additional areas, such as citizenship of the European Union and elections to the European parliament. In the covered areas, the assent of the European parliament is required for proposals.

The European Parliament. The Treaty of European Union provides the authority for the European parliament to represent the people of the member states (Article 189(1) EC).[26] Article 189 describes the role of Parliament thus: "acting jointly with the Council and the Commission, (Parliament) shall make regulations and issue directives, take decision, make recommendations or deliver opinions."[27]

In addition to working with the council and commission, Parliament also responds to citizen petitions. The *Rules of Procedures of the European Parliament* provides that individual citizens may, individually or as a group, use their right of petition to submit grievances to Parliament on matters that fall within the European Union's fields of activity and that affect the citizen(s) directly.[28] A parliamentary committee will review the petitions, may arrange hearings, and/or request the submission of documents or information from the commission.[29] The committee reports to Parliament every six months on the outcome of the outcomes of its investigations of petitions.

Once the investigations are completed, the president of Parliament informs the petitioners of its decisions.

Since 1979 members of the European parliament have been elected by general election every five years.[30] (Before 1979, the representatives were selected from among the members of national parliaments under the initial plan laid out by the founding treaties.) Article 190(4) of the Amsterdam Treaty provides for Parliament to develop uniform procedures to be followed in all member states or that the election procedures in the member states follow "principles common to all Member States."[31] As uniform procedures have not been provided for as of March 2002, the requirement for common principles is being followed. Article 19 (8b) of the EC Treaty further provides that "every citizen of the Union residing in a Member State of which he is not a national shall have the right to vote and to stand as a candidate in elections to the European parliament in the Member State in which he resides" (Directive 93/109/EC).[32]

The member states elect the members of the European parliament by direct universal suffrage under a system of proportional representation.[33] In the majority of the states, the entire country forms the electoral area, with elections held on a national basis.[34] In others the system differs, for example, in Belgium, Italy, and the United Kingdom, elections are held on a regional basis.[35]

Parliament's membership of 626 is fairly proportional in size to each member state's population, although the smaller countries are slightly overrepresented. The Treaty of Nice limits the number of European parliament members to 732, to be allocated between member states and candidate countries (after they become members).[36]

The members of the European parliament meet monthly in full session in Strasbourg to amend and vote on draft legislation and policy. Additional plenary sessions may be held in Brussels. Members do not sit in national delegations but in multinational political groups. In 2002, the largest party was the European People's Party (Christian Democrats) and European Democrats (one party), which held 232 of the 626 seats.[37] The second largest group was the Party of European Socialists, with 175 seats.[38]

Like the commission, Parliament also has budgetary influence because the budget must be accepted by Parliament before it can be adopted. Parliament has the authority, within certain limits, to amend the draft budget for noncompulsory items. In the case of compulsory expenditures, however, Parliament can propose modifications only by absolute majority. (The largest single item in the budget, agricultural expenditure, is compulsory.)

Parliament shares decision-making power over general action programs with the Council of Ministers. The president of the council is required to provide Parliament a report of council meetings. Parliament has the right to ask questions of the Council of Ministers in either written or oral form. On matters that require unanimity in the council, Parliament may be consulted; however, its opinions are not binding. Additionally, the foreign minister of the state holding the presidency of the council is required to report to the parliament at both the beginning and the end of that state's presidency.

In theory, the European parliament has some supervisory power over the commission and council. However, in reality, supervisory influence is limited. For example, although it has the right to be consulted by the member states when they are agreeing on a new president of the commission, Parliament was not consulted in 1994. Parliament also has the power to dismiss the entire commission on a vote of censure (by a two-thirds majority vote); it did so in 1999. Parliament cannot target individual commissioners.

The European Court of Justice. The role of the Court of Justice is to interpret community law and adjudicate disputes. The Court of Justice is made up of one judge from each member state.[39] Appointment is for a renewable term of six years, with half the court turning over every three years. The president of the court is chosen by the members of the court and serves a renewable term of three years. All judges are required to meet the qualifications necessary for appointment to the highest judicial offices in their home states, although they need not have served in such a capacity in their home states. The types of cases resolved before the court include disputes between and among member states, the European Union, and various institutions, individuals, and corporations. The court may also provide opinions on international agreements.

Although few of the early cases brought before the court involved environmental matters, the court generally found a basis for support for environmental concerns. For example, the 1980 legal decision in *Simmenthal SpA v Commission* provided that national governments must apply community law in full and allowed individuals to appeal when this is not done. A few years later this was put to the test in the 1991 *Francovich* case, which established that individuals could take a government to court for failure to properly implement directives related to the environment.[40]

The most common court procedure involving environmental matters is infringement. The commission can initiate infringement proceedings when a member state fails to fulfill its treaty obligations (under the Treaty of Rome, Art. 169). The member state is given a period of time to comply; this occurs during a mandatory round of consultation between the commission and the member state. If the member state does not comply, the commission may file suit. Cases involving the environment have increased substantially in recent years, resulting in over 220 cases brought before the court between 1997 and 2001 related to the environmental concerns. These cases have dealt with issues of community law implementation such as the failure of Ireland to conform to community directives regarding the treatment of animals used in experiments (359/99, referring to case number 359 of the year 1999), the failure of Spain to conform to ambient air quality assessment and management directives (417/99), and the failure of Germany to provide listings as required by directives on the conservation of natural habitats (71/99).[41] The court ruled against the defendants in each of these cases.

The European Council. The European Council, also known as the EU heads of state or government, is made up of the heads of government of the member states. Although early treaties did not discuss this group, meetings in the 1960s, 1970s, and 1980s internally established the role of the European Council in the community. Over the years, European Council exchanges became more regular and helped give direction to, and resolve problems in, the community. The role of the European Council was finally recognized in Article 4 of the Treaty on European Union (however, the treaty does not provide for the council as a legal institution of the European Community). The role of the council under Article 4 is to "provide the Union with the necessary impetus for its development and define the general political guidelines thereof."[42]

The membership of the European Council includes the heads of state of the fifteen member states and the president of the European Commission. The ministers of foreign affairs of the member states assist and collaborate at European Council meetings, which generally occur twice a year.

Other official bodies are provided for in the treaties and serve to advise the commission, council, and parliament and to make comments on legislation. These include the European Economic and Social Committee and the Committee of the Regions, which are described below. In addition, the European ombudsman is provided for in the Treaty of European Union to serve as a means for citizens to seek redress if they feel that they have been harmed by any governmental body. The role of the ombudsman is also described below.

The European Economic and Social Committee. The role of the European Economic and Social Committee is to advise the commission, the Council of Ministers, and the European parliament on matters provided for in the Treaty of Rome, which specifically include agriculture and transport.[43] The committee may also issue opinions on its own initiative.[44] Between six and twenty-four members represent each member state, depending on the size of the state, for a total of 222 members. (The number of representatives from each country in the Economic and Social Committee is provided in table 2.2 below.) The Treaty of Nice provides that the number of representatives on the Economic and Social Committee not exceed 350.[45] This action was taken to prepare for the addition of new member states to the EU in the future. Committee members, who are not employees of the union, generally continue in their regular line of work. Members are proposed by national governments and appointed by the Council of Ministers. Terms are for four years and are renewable. A chairperson, who is elected by the committee and serves for a term of two years, leads the committee. The committee meets monthly. The Treaty on European Union, Articles 193 and 194, provides the following for the Economic and Social Committee:

> It shall have advisory status.
>
> The Committee shall consist of representatives of the various categories of economic and social activity, in particular, representatives of producers, farmers, carriers, workers, dealers, craftsmen, professional occupations and representatives of the general public.
>
> The members of the Committee may not be bound by any mandatory instructions. They shall be completely independent in the performance of their duties, in the general interest of the Community.[46]

Committee of the Regions. The Committee of the Regions was created by the Treaty on European Union to represent local and regional bodies. The major goal of the committee is to bring the union closer to the public by acting as a representative of the people. The Treaty on European Union provides for the Committee of the Regions to act in the interests of the community:

> A Committee consisting of representatives of regional and local bodies, hereinafter referred to as "the Committee of the Regions," is hereby established with advisory status.... The members of the Committee may not be bound by any mandatory instructions. They shall be completely independent in the performance of their duties, in the general interest of the Community. (Article 198A)[47]

The Treaty mandated that the Committee of the Regions be consulted on five areas likely to have repercussions at the regional or local level: economic and social cohesion, trans-European infrastructure networks, health, education, and culture. The Treaty of Amsterdam extended the areas on which the committee is to be consulted to include the environment, social policy, employment, vocational training, and transport.[48] The Committee of

Table 2.2.
Number of Representatives in the Economic and Social Committee from each Member State

Member State	*Number of representatives in the Economic and Social Committee*
France	24
Germany	24
Italy	24
United Kingdom	24
Spain	21
Austria	12
Belgium	12
Greece	12
Netherlands	12
Portugal	12
Sweden	12
Denmark	9
Finland	9
Ireland	9
Luxembourg	6

Source: *Euroconfidential, The Rome, Maastricht, and Amsterdam Treaties: Comparative Texts* (Belgium: Euroconfidential, 1999), 179.

the Regions also has six internal commissions that specialize in various fields, one of which is a Commission for Sustainable Development.[49] Furthermore, the Committee of the Regions may issue opinions when it determines that a region's interests are involved. According to the text of Article 198c of the Treaty on European Union, as amended by Article 265 of the Amsterdam Treaty, the Committee of the Regions may be consulted in particular instances and may issue opinions on its own initiative when appropriate:

> The Committee of the Regions shall be consulted by the Council or by the Commission where this Treaty so provides and in all other cases, in particular those which concern cross-border

> cooperation, in which one of these two institutions considers it appropriate.
>
> Where the Economic and Social Committee is consulted pursuant to Article 262, the Committee of the Regions shall be informed by the Council or the Commission of the request for an opinion. Where it considers that specific regional interests are involved, the Committee of the Regions may issue an opinion on the matter.
>
> The Committee of the Regions may be consulted by the European Parliament.
>
> It may issue an opinion on its own initiative in cases in which it considers such action appropriate.
>
> The opinion of the Committee, together with a record of the proceedings, shall be forwarded to the Council and to the Commission.[50]

The Committee of the Regions has 222 members and 222 alternates who are nominated by the member states and appointed for four-year terms by the council. The members of the committee are based in their home regions or local areas and meet for five plenary sessions in Brussels every year and may also attend commission meetings.[51] The number of representatives from each country is provided for in the Treaty on European Union and is the same as that for the Economic and Social Committee. Table 2.3 provides the number of representatives from each member state in the Committee of the Regions. Figure 2.5 provides an excerpt from an opinion of the Committee of the Regions on council on public participation in the development of plans relating to the environment.

The European Ombudsman

The European ombudsman was established by the Treaty on European Union and provides an avenue for citizens to seek redress if the citizen is a victim of an act of "maladministration" by any community institution or body, except the Court of Justice and the Court of First Instance.[52] The European parliament elects the ombudsman for renewable five-year terms.

The ombudsman investigates inquiries based on citizen complaints. He or she can also launch his or her own inquiries. Member states and community institutions and bodies are required to supply information as requested by the ombudsman. The role of the ombudsman is to try to find a solution agreeable to the parties and/or make recommendations to solve the

Table 2.3.
Number of Representatives in the Committee of the Regions from Each Member State

Member State	*Number of representatives in the Committee of the Regions*
France	24
Germany	24
Italy	24
United Kingdom	24
Spain	21
Austria	12
Belgium	12
Greece	12
Netherlands	12
Portugal	12
Sweden	12
Denmark	9
Finland	9
Ireland	9
Luxembourg	6

Source: *Euroconfidentiel, The Rome, Moastnicht, and Amsterdam Treaties: Comparative texts* (Belgium: Euroconfidential, 1999), 179.

case. He or she can also make a special report to the European parliament regarding the matter.

Cases brought before the ombudsman have covered a wide variety of matters, such as, lack of access to information, administrative delay, and recruitment of staff by community institutions. Between 1995 and 2002 the ombudsman reviewed thirty-six cases in the environment area. All but one of these complaints were complaints against the European Commission; the other was against the European Investment Bank. In seventeen of the cases the ombudsman found no instances of maladministration.[53] The other cases were settled by the European Commission responding to the complaints. Figures 2.6 and 2.7 describe instances of complaints brought against the

Figure 2.5.
Selected Excerpts from the Opinion of the Committee of the Regions on Amendments to Council Directives 85/337/EEC and 96/61/EC on Public Participation in the Development of Plans Relating to the Environment

Council Directive 85/337/EEC on the assessment of the effects of certain public and private projects on the environment, as amended by Council Directive 97/11/EC provides for environmental impact assessments. The Council specifies that,

> "consent for public and private projects which are likely to have significant effects on the environment should be granted only after prior assessment of the likely significant environmental effects of these projects has been carried out; whereas this assessment must be conducted on the basis of the appropriate information supplied by the developer, which may be supplemented by the authorities and by the people who may be concerned by the project in question... projects belonging to certain types have significant effects on the environment and these projects must as a rule be subject to systematic assessment..."
>
> Whereas the effects of a project on the environment must be assessed in order to take account of concerns to protect human health, to contribute by means of a better environment to the quality of life, to ensure maintenance of the diversity of species and to maintain the reproductive capacity of the ecosystem as a basic resource for life."[54]

Council Directive 96/61/EC, with the amendments proposed by the council, establishes a scheme for greenhouse gas emission allowance trading within the community. "The proposal, based on Article 175(1) of the Treaty, places direct emissions of the greenhouse gases covered by the Kyoto Protocol within a regulatory framework."[55] The directive provides that member states will grant permits and allocate allowances for greenhouse gas emissions to companies. The permits may be traded between companies.

The Committee of the Regions on these Directives commended the directives as a step toward improving citizen participation. It pointed out, however, that citizen participation should be encouraged at all areas of local and public planning. The committee held that public access to information and opportunities to be involved in decision making was vital. They also held the view that environmental organizations should not be allowed to override the desires of other concerned citizens.

> The broad definition of "the public concerned," as including environmental non-governmental organizations... is welcomed by the Committee of the Regions... but in practice it is likely to increase the

Figure 2.5. ***(Cont'd)***
Selected Excerpts from the Opinion of the Committee of the Regions on Amendments to Council Directives 85/337/EEC and 96/61/EC on Public Participation in the Development of Plans Relating to the Environment

extent to which environmental interest and pressure groups are able to delay the implementation of necessary development projects, even where every effort has been made to avoid, minimize, mitigate or compensate for environmental impacts of that development... A look should be taken at the definition of "the public concerned" so that consumers/users associations, and professional associations of all types and levels could be expressly included alongside environmental NGOs.

Obviously there needs to be a careful balance struck here between executive action and scrutiny... It will be important for Member States, in determining which associations (NGOs, non-profit organizations of benefit to society, sectoral associations, consumer and user associations, civil protection volunteer associations, welfare associations, etc.) fulfill the requirements for having a legitimate interest, to consider this point."[56]

Among its recommendations the Committee of the Regions calls for the European Commission to collect and disseminate information regarding best practices for providing and encouraging citizens to become involved in the development of plans and strategies for their areas. The committee also suggested that parties relevant to decision making should not be limited to environmental NGOs, but may also include other organizations such as consumer groups.

Source: "Opinion of the Committee of the Regions on the 'Proposal from the Commission for a Directive Providing for Public Participation in Respect of the Drawing Up of Certain Plans and Programmes Relating the Environment and Amending Council Directives 85/337/EEC and 96/61/EC,'" Official Journal C 357/58 14.12.2001. Accessed 26 March 2002. Available from < http://europa.eu.int/eur-lex/pri/en/oj/dat/2001/c_357200112/4en00580060.pdf >

European Commission in the environment field in which the ombudsman found the European Commission in violation.

Summary of EU Governing Structure

The commission, the Council of Ministers, and the European parliament are the dominant institutional actors in the European Union's legislative process.

The EU governing system is designed to include representation from a variety of groups within the system, such as the European Council, the Economic and Social Committee, and the Committee of the Regions, in order to better represent the people of the union.

Every group, formal or informal, governmental or nongovernmental organization (NGO), is benefited by the availability of good information. That is the primary duty of the Europe Environment Agency.

THE EUROPEAN ENVIRONMENT AGENCY

In order to generate better policy decisions on the environment and better implementation of policy, the Council of Ministers created the European Environment Agency (EEA) (adopted 7 May, 1990).[57] The agency's mission is "to support sustainable development and to help achieve significant and mea-

Figure 2.6.
Decision of the European Ombudsman on Complaints 271/2000(IJH)JMA and 277/2000(IJH)JMA against the European Commission

In 2000, a complaint was lodged on behalf of the environmental organization, Friends of the Earth, against the European Commission "for its refusal to supply copies of two studies commissioned by the institution to an independent consultant. These reports concerned compliance of the UK and Gibraltar with the Community Directives on waste (Directive 75/442/EEC) and hazardous waste (Directive 92/689/EEC), as well as with the Habitats Directive (92/43/EEC)."

The commission responded that "some of the information contained in the report was covered by the exception involving the protection of public interest (inspections and investigations) provided for under the Code of Conduct concerning Public Access to Commission documents (Decision 94/90/EEC)."

In response to the commission's statement, the complainant alleged that the "Commission's decision to reject his requests for access were unlawful because...the public interest exception should not apply to an independent and objective third party document. Independent reports cannot be considered as internal Commission documents, and therefore the exceptions provided for in Decision 94/90/EEC should not apply to this type of documents."

The ombudsman found that "the Commission had wrongly refused access to Commission documents (and the Commission should reconsider its decisions in this case).... Such action constituted an instance of maladministration."

The commission accepted the ombudsman's recommendations and provided the complainant with a full copy of the requested studies.

Source: Decision of the European Ombudsman on complaints 271/2000/(IJH)JMA and 277/2000/(IJH)JMA against the European Commission, <http://www.euro-ombudsman.-eu.int/decision/en/000271.htm> 24 March 2002.

Figure 2.7.
Decision of the European Ombudsman on Complaint 493/2000/ME against the European Commission

In 2000, a complaint was made to the European Ombudsman on behalf of "Västkustbanans Framtid" regarding the Environmental Impact Assessment made for the Swedish train connection "Västkustbanans" and the classification of the "Skälderviken" area in Sweden under the "Habitats Directive." The president of "Västkustbanans Framtid" alleged that "Sweden had failed to comply with Directive 85/337/EEC and Directive 92/43/EEC on the conservation of natural habitats and of wild fauna and flora...The Environmental Impact Assessment was inadequate as it did not cover all relevant factors.... The Commission should not have accepted the EIA."

The commission responded that "the submissions made by the complainant did not demonstrate that the assessments were inadequate.... Directive 85/337/EEC regulates the procedure rather than the substance or quality of the assessment."

In his decision regarding the part of the complaint involving the Environmental Impact Assessment, the ombudsman noted, "Directive 85/337/EEC requires an exercise of judgment by the Member State as to what information is to be provided.... The ombudsman's inquiry revealed no evidence to show that the Commission was not entitled to take the view that the Member State had complied with it obligation under Directive 85/337/EEC. " However, in his conclusion, the Ombudsman included a critical remark: It is good administrative behavior to state the reasons for a decision. The reasons given should be adequate, clear and sufficient. In the present case, the Ombudsman finds that the Commission failed to state the reasons for its decision.

In response to the allegations regarding the Skälderviken area, the commission opened a new complaint file and noted that it would inform the complainant of the action to be taken. Thus, the ombudsman determined that there was no maladministration regarding the part of the complaint regarding the classification of the Skälderviken area as a Natura-site under Directive 92/43/EEC.

Source: Decision of the European Ombudsman on Complaint 1561/2000/PB against the European Commission, < http://www.euro-ombudsman.eu.int/decision/en/000493.htm > 24 March 2002.

surable improvement in Europe's environment through the provision of timely, targeted, relevant, and reliable information to policy-making agents and the public."[58] The agency has as its main objective, under Regulation 1210/90, the provision of objective, reliable technical and scientific information regarding environmental concerns to the community and member states. The agency has no regulatory or enforcement authority per se. Its goal is to help member states implement the measures required to protect their environment and to evaluate the results of such measures. A second goal is to provide information to the public regarding the state of the environment. The agency has the unique property of being open to countries that are not members of the European Union but share overlapping environmental concerns. In 2001, non-EU members included Iceland, Liechtenstein, and Norway.

The agency is supervised by a management board, which is made up of one representative from each member state and two representatives designated by the commission. An executive director, who serves a renewable term of three years, heads the agency. The executive director is assisted by a deputy director and representatives from the member states, who are generally associated with their respective state's national environment agency. The European parliament chooses two other members, who are identified in Regulation No. 1210/90, which established the European Environment Agency, as "scientific personalities, particularly qualified in the field of environmental protection."[59] Members selected by the European Commission to work with the agency include the director, Directorate B DG Environment and the director of the Environment Institute Joint Research Center.[60] To perform its mission the agency has allocated to it a multiannual budget, with funding provided by the EU. Contracts with third parties account for almost half the budgeted expenditures. The support staff numbers approximately sixty.[61] The agency is headquartered in Copenhagen.

The agency is assisted in its work by the European Information and Observation Network (EIONET), a subsection of the agency. EIONET helps the agency gather information, identify issues, and produce information. EIONET considers a variety of environmental matters, including air, water, waste management, land use, and protection of flora and fauna. The agency works with EIONET to gather information in member states. The agency also works with other bodies and international organizations so that duplication of efforts can be avoided.

During its first two years, the agency began its work by recording, collating and assessing data.[62] In 2001, the agency's focus was to:

—Provide the Union, Member States, and third countries with objective information for drawing up and implementing effective environmental protection policies
—Supply the technical, scientific and economic information required for laying down, preparing and implementing measures and laws related to environmental protection
—Develop forecasting techniques to enable appropriate and timely preventive measures to be taken

—Ensure that European environmental data are incorporated into international environmental programmes.[63]

The agency is continuing to work toward building EIONET, identifying emerging environmental issues, supporting the framing and development of environmental policies and the evaluation of environmental policies, and developing an integrated environmental monitoring and reporting process.[64]

The Three Pillars of the European Environment Agency

The agency structures its work around three so called "pillars": the Reference Center, networking, and DPSIR. Each of these provides part of the structure required for providing support for decision making and policy action. The Reference Center provides data and information. Networking includes interaction with various environmental agencies, both within and outside the EU. Projects with agencies outside the EU include, for example, coproducing a joint statement on chemicals in the environment with UNEP. Projects with EU entities include establishing an Internet-based forum for document exchange.

DPSIR stands for the five elements that form the framework for reporting on environmental issues (the Drivers, industry and transport; the Pressures, polluting and emissions; the State, air, water, soil quality; responses, clean production, public transport; the Impact, ill health, biodiversity loss, economic damage; and the Responses, regulation, taxes, information).[65] The role of the EEA under the DPSIR includes providing information on the elements of the DPSIR, their interconnections, and the effectiveness of responses to the various environmental pressures and their impacts.

Work done by the agency has included establishing an annual reporting system in thematic areas (such as urban air quality) and developing and updating technical manuals such as the *Guidance Report for Air Quality Management Assessment.* The agency also generates a variety of publications including topic reports, such as *Towards a European Habitat Classification,* as well as periodical reports, such as *Europe's Environment: The Second Assessment.*[66] Figure 2.8 provides a list of selected European Environment Agency publications.

Besides using the EEA to provide input on environmental issues, the EU also provides a role for the public in environmental policy formation. As we see below, public input and action is taken both formally and informally.

THE ROLE OF THE PUBLIC IN EU POLICY MAKING

The role of the public in environmental protection is evident throughout the EU, as environmental action groups and related organizations bring public

Figure 2.8.
Selected Publications of the European Environment Agency

A Checklist for State of the Environment Reporting, Technical Report No. 13 (01 January 1999).
Air Pollution in Europe 1997, Environmental Monograph No. 4 (21 July 1999).
Analysis and Comparison of National and EU-wide Projections of Greenhouse Gas Emissions, Topic Report No. 1/2002 (23 August 2002).
Annual European Community Greenhouse Gas Inventory 1990–1999— Submission to the Secretariat of the UNFCCC, Technical Report No. 60 (29 April 2002)
Cloudy Crystal Balls: An Assessment of Recent European and Global Scenario Studies and Models, Environmental Issue Report No. 17 (15 December 2000).
Computer-based Models in Integrated Environmental Assessment, Technical Report No. 14 (25 October 1999).
EIONET Noise Newsletter, No. 3, January 2000 (27 January 2000).
Environmental Signals 2001, Environmental Assessment Report No. 8 (29 May 2001).
Genetically Modified Organisms (GMOs): The Significance of Gene Flow Through Pollen Transfer, Environmental Issue Report No. 28 (21 March 2002).
Household and Municipal Waste: Comparability of Data in EEA Member Countries, Topic Report No. 3/2000 (10 October 2000).
Sustainable Water Use in Europe—Part 2: Demand Management, Environmental Issue Report No.19 (5 April 2001).

Source: European Environment Agency Website, accessed 21 October 2002, http://reports.eea.eu.int/index_table?sort=Language#English.

attention to environmental issues.[67] Europeans are very aware of and concerned with environmental issues.[68] Increasingly, environmental action by the public is being recognized under EU law. EU law requires environmental data to be made available to the public. As noted above, the European Environment Agency has this assignment as one of its objectives.

Role in Policy Formation

Indirectly and informally, the public plays an important role in EU environmental policy formation. As previously mentioned, there are public hearing and notification requirements in the environmental assessment process. In addition, every EU country has local environmental interest groups that (to varying degrees) participate in environmental policy formation within their countries. From the Royal Society for the Protection of Birds in Great Britain to Friends of the Earth in Finland, all of these many organizations have an impact on environmental regulatory thinking, which, ultimately, impacts environmental policy formally in the EU.

The EU has made strong efforts to encourage the public to participate and provide input into the policy-making process. The EU is a signatory to the UN Convention on Access to Information, Public Participation in Decision-Making and Access to Justice in Environmental Matters, adopted in Aarhus, Denmark, in 1998. The convention provides principal rules for the promotion of citizen involvement in environmental issues, based on three pillars: (1) the right of access to environmental information, (2) the right to participate in decision-making processes, and (3) access to justice for the public.[69] Building on this theme, in 2002 the EU developed a set of minimum standards for consultation of interested parties to "get citizens more actively involved in achieving the Union's objectives and to offer them a structured channel for feedback, criticism, and protest".[70] The minimum standards set down by the European Commission are provided in Figure 2.9.

Citizens of the EU have access to consultations, discussions, and other tools to provide an avenue for citizens to actively take part in the policy-making process by using the EU Internet site, Your Voice in Europe, which provides access in all official EU languages.[71] Your Voice in Europe was developed as part of the European Commission's Interactive Policy Making Initiative (IPM-C(2001)1014) to provide for the evaluation of existing policies and for open consultations on new initiatives.[72]

Lobbying

While the public and environmental organizations in the United States often make their wishes known through lobbying directly to the elected representatives, lobbying in the EU is a complex task because the institutions of the union have different priorities and make their decisions in different ways. The commission itself does not concern itself with any individual state or organization, particularly those from outside the EU. Therefore, non-Europeans either join in a common cause with a community industry or act through a European intermediary organization.[73] The timing of the lobbying effort at the commission level is also central to any lobbying effort. Once the commission finalizes the preliminary draft of a proposal, only minor changes are possible. In the European parliament, the level at which lobbying takes place is the committee, where the direction of legislation is determined. Once Parliament presents its report for the First Reading or Opinion of Parliament, possibilities for change are minimal. Lobbying of the Council of Ministers is not possible. The most effective method of lobbying appears to be to work at the level of the member states.

Figure 2.9.
Minimum Standards for Consultation of Interested Parties by the Commission

The basic principals for consultation by the Commission include the following provisions:

To encourage more involvement of interested parties through a more transparent consultation process, which will enhance the Commission's accountability.

To provide general principles and standards for consultation that help the Commission to rationalize its consultation procedures, and to carry them out in a meaningful and systematic way.

To build a framework for consultation that is coherent, yet flexible enough to take account of the specific requirements of all the diverse interests, and of the need to design appropriate consultation strategies for each policy proposal.

To promote mutual learning and exchange of good practices within the Commission.[74]

As part of the effort to encourage and provide for participation on the part of the public, the Directorate General of the Environment offers regularly open consultations to members of the public on various environmental concerns. In 2003, open consultations included discussions on INSPIRE (Infrastructure for Spatial Information in Europe) and on draft chemicals legislation. Interested parties may register their comments online. Closed consultations with specific stakeholders or other interested parties are also held. Policy activities discussed in the consultations have included various environmental issues, including NATURA 2000, waste, pesticides, and natural and technological hazards.

CONCLUSION

In this chapter, we have discussed the foundational laws and treaties affecting environmental policy making in the EU. The organizational structure, as we have seen, also influences policy making. The next chapter focuses more directly on environmental policy and its implementation in the EU. We review the history of environmental policy development in the EU. We also examine the environmental action programs that have been developed by the EU to guide environmental action.

CHAPTER 3

THE EUROPEAN ENVIRONMENT

INTRODUCTION

Understanding European environmental policy entails an examination of many things. In the last chapter we summarized the major legal and institutional components of the European Union and the historical development of the treaties that created those institutions.

In this chapter, we begin with a very brief description of the environment of Europe and then summarize the major environmental policy initiatives that have come out of the European Union. We conclude with a brief overview of the European Environment Agency.

The European environment is quite diverse, ranging from the arid lands in the southern states to near-Arctic conditions in the northern states. It contains large, densely populated urban areas as well as areas with virtually no population. Woodlands and farmlands, coastal and inland areas, rivers and deserts are all part of the European environment.

The long history of human habitation in Europe has led to several environmental problems. Many waterways have been polluted by industrial and agricultural runoff. The air has suffered the effects of industrial pollution, automobile exhaust, and electricity generation. Northern forests have suffered from acid rain, and the southern areas have experienced desertification from intensive farming activities. Preservation of natural habitats is an issue everywhere.

The long history of human habitation has had other impacts on the European environment as well. For example, in Europe wilderness management means something different than it does in other parts of the world. Human impact is so much a part of most of the European landscape that lands have to be managed in ways that make human activity part of the ecosystem—even if people are removed.[1]

Because Europe is made up of relatively small states, the actions of one state often reach beyond its own boundaries to affect another. The nations of Europe have recognized that although each has sovereignty within its own borders, measures must be taken to deal with these spillover effects. Thus the European Union has made protection of the environment one of its primary goals. Figure 3.1 provides a description of the objectives of the European Commission regarding the environment.

Figure 3.1.
Objectives of the European Commission on the Environment

To maintain and improve the quality of life through a high level of protection of our natural resources, effective risk assessment and management and the timely implementation of Community legislation.

To foster resource-efficiency in production, consumption and waste-disposal measures.

To integrate environmental concerns in to other EU policy areas.

To promote growth in the EU that takes account of the economic, social and environmental needs both of our citizens and of future generations.

To address the global challenges facing us notably combating climate change and the international conservation of bio-diversity.

To ensure that all policies and measures in the above areas are based on a multi-sectoral approach, involve all stakeholders in the process and are communicated in an effective way.

Source: European Commission, "Environment: Overall Objectives," accessed 30 October 2001. http://www.europa.eu.int.comm/budte5/abb/environment_en.htm.

HISTORY OF ENVIRONMENTAL POLICY DEVELOPMENT

The original treaties establishing the European Communities did not contain any references to the environment (although the Treaty of Rome did contain general references to the quality of life in Articles 2 and 36). The Paris Conference of the Heads of States and Governments, held in conjunction with the Stockholm Conference on the Human Environment (1972), formed the impetus for including environmental considerations in Community policy with a declaration stating, "As benefits to the genius of Europe, particular attention will be given to intangible values and to protecting the environment, so that progress may really be put at the service of mankind."[2] (The Stockholm Conference is often noted as the point at which the nations of the world officially recognized environmental concerns such as global warming.)

The Single European Act formalized and made explicit strong involvement in the environmental area on the part of the members of the European Community. Previous to that time, economic development was the main priority of community policy. Article 130r of the Single European Act explicitly makes environmental protection a component of other Community policies. As noted above, Articles 100a, and 130r to 130t of the Treaty on European Union and Article 2 of the Treaty of Amsterdam determine the objectives of EU environmental policy, as well as the methods for instituting and implementing those objectives. The basis for environmental measures in the community are based on Article 100, which mandates the council to harmonize national measures regarding products that may distort competition, and Article 235, which empowers the council to take measures that contribute to the promotion of the basic goals of the various community treaties.

Environmental concerns have resulted in over two hundred pieces of EU legislation, covering all areas of the environment. Environmental legislation is most prolific in the areas of water pollution, atmospheric pollution, noise, chemical products, waste disposal, and nature protection.[3] Water quality standards have been set and include regulations on the discharge of toxic substances. The EU has also participated in efforts to reduce pollution in international waters. To improve air quality, the EU has adopted directives to deal with pollution from power stations and motor vehicle exhaust. Maximum noise levels have been established on vehicles and various types of motorized machinery. Numerous directives have regulated both chemical products and waste disposal. The EU is a member of several conventions for the conservation of wildlife, birds and wetlands, and several conservation-related directives have been adopted. Figure 3.2 provides a list of representative environmental conventions to which the EU is a signatory.

One of the key instruments used in EU environmental policy is the environmental impact statement. Effective in 1988, the directive on environmental impact assessments requires all major projects that may significantly impact the environment to perform a comprehensive assessment of potential environmental impacts. A declaration on environmental impact assessments is also included in the Final Act of the Intergovernmental Conference, which drafted the Treaty of Amsterdam. This declaration (Directive 97/11/EC, amending Directive 85/337/EEC) calls for the commission to prepare environmental impact assessment studies when making proposals for projects that may impact the environment.

The directive on environmental impact assessments provides a procedure for assessing the potential impact of projects on the environment.[4] The directive requires that specific criteria be considered, including the use of natural resources and the environmental sensitivity of the area. The United States was a leader in creating the environmental impact statement (EIS) process. Many of the requirements of the directive on environmental assessments are similar to the system developed in the United States. (The EIS process in the United States has not lived up to its potential.[5] We might hope

Figure 3.2.
Selected Environmental Conventions to Which the EU Is a Signatory

Barcelona Convention for the Protection of the Marine Environment and the Coastal Region of the Mediterranean, 1976

Bern Convention on the Conservation of European Wildlife and Natural Habitats, 1971

Bonn Convention on the Protection of Migratory Animals, 1983

Convention for the Protection of the Rhine, 1976

Convention on Biodiversity, 2000

Helsinki Convention on the Protection of the Marine Environment of the Baltic Sea, 1992

Stockholm Convention on Persistent Organic Pollutants, 2000.

The Cartagena Biosafety Protocol, 2000

The Kyoto Protocol on Climate Change, 1997

The Rio World Summit: The Framework Convention on Climate Change, 1992

Source: European Commission, "Europe's Role in Worldwide Action," in *Environment for Europeans* No. 10 (April 2002). Accessed 23 October 2002. Available from http://europa.eu.int/comm/environment/news/efe/index.htm.

that the EU has a better experience.) As in the United States, public participation is to be solicited. The goal is to identify potential impacts so that decision makers can take whatever measures are necessary to prevent or minimize any potential negative effects. Results of the assessment must be made available to the public before approval can be granted.

In recent years several additional steps have been taken toward strengthening the environmental focus in the union. In 1990 the council adopted a directive (90/313) that provides public access to environmental data. The directive on the freedom of access to information on the environment, effective in 1993, provides public access to environmental information held by public authorities. National authorities are required to make information on the environment available upon request; however, the directive also allows member states to determine the conditions under which this information may be provided.

In the member states of the EU, the implementation of community legislation regarding the environment is the responsibility of the national government. In most EU countries the executive organization covering environmental affairs is a public body directly or indirectly responsible to the state.[6] The community sometimes introduces common policies, with the community as the responsible party. In other instances the community coordinates national policies, in which case the member state retains responsibility. Thus environmental policy is the not simply the policy of the individual state but is policy that has been determined at the community level.

Challenges to Environmental Policy Making

In spite of the large number of legislative actions supporting environmental protection, problems still exist in many areas. Some of the problems associated with implementing environmental measures are discussed below.

Subsidiarity Issues. One significant impediment to the progress of environmental policy in the European Union has been the issue of subsidiarity. As described in chapter 2, the principle of subsidiarity is defined in Article 3b of the Maastricht Treaty thus:

> In areas which do not fall within its exclusive competence, the Community shall take action, in accordance with the principle of subsidiarity, only in and insofar as the objectives of the proposed action cannot be sufficiently achieved by the Member States and can therefore, by reason of the scale or effects of the proposed action, be better achieved by the Community.[7]

Subsidiarity identifies where, in areas in which the EC does not have sole power, it can still take action on behalf of larger community interests.[8] Recall that Article 3b also provides that "the Community should act only where action at the Community level is the most efficient approach."[9]

Both the Committee of the Regions and the European Policy Center have issued opinions regarding the issues of subsidiarity as described in the Treaty on European Union and its revisions in the Amsterdam Treaty. Figure 3.3 provides excerpts from the Committee of the Regions' "Opinion on the Revision of the Treaty on European Union," in which they point out that the Maastricht Treaty is the "first text that provides for regional and local participation in the drawing up of Union policies."[10] The committee's opinion supports the principle of subsidiarity, but notes that the appropriate applications of the principle need to be clarified. Figure 3.4 provides excerpts from the European Policy Center's working paper, "Beyond the Delimitation of Competences: Implementing Subsidiarity," in which it describes the role of subsidiarity in the union.[11] The center's paper states that a balance between appropriate level of decentralization of decision making and the use of the subsidiarity principle for decision making at the union level needs to be achieved.

Implementation and Enforcement. Even though sustainability has become a major objective in the EU with the adoption of the Treaty of Maastricht, some of the member nations have been more aggressive than others in attaining that goal. As a result, regulatory processes and strategies for their implementation vary across the union. Research indicates that the Netherlands and Germany have been strong advocates for the environment. However, other countries, such as Italy and Greece, have not been as aggressive in implementing European Union directives.[12]

Because of this variation across the union, there was a trend in the 1990s to move policy-making power to the community level and away from the member states.[13] This move allowed environmental action at the community level to have a broader scope. However, policy is still required to consider the diversity of situations throughout the community. According to Article 130 of the Treaty on European Union, community policy on the environment is to aim for a high level of protection, leading to sustainability yet taking into consideration relevant differences among the member states.[14]

Another reason for the irregular application and uneven success of environmental initiatives is lack of enforcement. This may result because member states may lack mechanisms to enforce environmental legislation. In addition, the commission itself suffers from insufficient staff so that violations are often not discovered or are recognized only when complaints are filed.

At times policies also appear to be contradictory. For example, the implementation of the European Union's Common Agricultural Policy has led to increased agricultural pollution, partially as a result of increased use of fertilizers and pesticides. However, nitrogen emissions from fertilizers lead to soil acidification and depletion of the ozone layer, causing environmental harm. Thus it appears that there may be a trade-off between environmental and market concerns.

While there are still challenges for environmental policy making in the EU, much progress has been made. The EU has made clear its goals and strategies for the environment in a series of environmental action programs, which we discuss in the next section.

ENVIRONMENTAL ACTION PROGRAMS IN THE EUROPEAN UNION

The EU's Environmental Action Programs are periodic plans that establish the community's strategy for the environment. The first environmental action program covered the period 1973 to 1976; the most recent, the Sixth Environmental Action Program, covers 2001 to 2010. The action programs are important because they define the priorities for the period covered by the

Figure 3.3.
Selected Excerpts from the "Opinion of the Committee of the Regions on the Revision of the Treaty on European Union" Related to the Issue of Subsidiarity

Excerpts from the "Opinion of the Committee of the Regions"

> By declaring subsidiarity to be a fundamental principle, the Maastricht Treaty limits Union activities to fields where efficiency requires supranational action; indeed it defines the Union, under the second paragraph of Article A, as one "in which decisions are taken as closely as possible to the citizen."

The Committee of the Regions provides the following proposal related to the principle of subsidiarity:

> The principle of subsidiarity implies that the public authorities do not take action when this can be done adequately and effectively by citizens. The principle also introduces the concept of gradation, i.e., higher levels of government act only when lower levels cannot do so satisfactorily. Subsidiarity in general, and subsidiarity within the process of European integration in particular, strengthens:

- ➢ democratic legitimacy, insomuch as it avoids the creation of an excessively centralized European power disconnected from the problems of ordinary citizens, the closeness of the Union to its citizens being one of the basic components of this legitimacy,
- ➢ transparency, since it encourages a clear-cut allocation of functions between various levels of government, making it easier for the citizen to identify areas of action appropriate to each level,
- ➢ efficiency, since it presupposes that powers are exercised at the most appropriate level of government.

> The Committee of the Regions... thus warmly welcomes the inclusion of the principle of subsidiarity in the Maastricht Treaty. It nevertheless regret that the concrete formulation of the principle of subsidiarity in Article 3b of the EC Treaty does no more than lay down a criterion for the exercise of shared powers between the Union and the Member States.
>
> The Committee of the Regions believes that the principle of subsidiarity needs to be looked at both in terms of its formulation in the Treaty and in terms of its applications.... We believe in particular that the Committee of the Regions must be more deeply involved in monitoring application of the principle of subsidiarity and must be brought into the heart of the work done by the Commission in this area.

Source: Committee of the Regions, "Opinion of the Committee of the Regions on the Revision of the Treaty on European Union" (Brussels, 20 April 1995). <http://europa.-en.int/en/agenda/igc-home/eu-doc/regions/crf_en.html> 25 March 2002.

Figure 3.4.
Selected Excerpts from the Working Paper of the European Policy Center, "Beyond the Delimitation of Competences: Implementing Subsidiarity"

The European Policy Center's comments on the principles of subsidiarity include the following:

> At the core of subsidiarity is the challenge to strike a delicate balance between freedom and efficiency. Higher levels of public authority should be involved only when their interventions deemed necessary for achieving objectives that would be out of reach for lower levels. Action should be taken as close as possible to citizens and left to private actors when the support of public structures is not required.
>
> [S]ubsidiarity and transparency, as basic requirements for legitimate decision-making, appear so interlinked that their separation, when dealing with the former, would be artificial.... Subsidiarity and transparency are complementary requirements for decision-making to be understood and for decisions to be accepted by the public.

The recommendations of the European Policy Center include the following:

> The key to improving the distribution of powers in the Union is to enhance the implementation of subsidiarity, bearing in mind the need to adjust to circumstances and priorities evolving at an accelerating pace.
>
> It should be acknowledged that subsidiarity does not necessarily mean decentralization and that, therefore, its proper application could also lead to more powers being entrusted to the Union, where Member States' action proves to be inadequate.

Note: The European Policy Center was founded in 1995 to promote policy discussions on the development of the union. It regularly publishes position papers and conference reports on EU events; it also publishes *Challenge Europe*, an online journal. In 2000, about 10 percent of its funding came from the European Commission for work on European integration. The remainder came from member and sponsors. Its members include corporations, diplomatic members, professional associations, nongovernmental organizations, foundations, religious, and intergovernmental organizations.

Source: European Policy Center, "Beyond the Delimitation of Competences: Implementing Subsidiarity: *The Europe We Need* Working Paper," 25 September 2001, <http://europa.eu.int/futurum/documents/other/oth250901_en.pdf> 24 March 2002.

program and set guidelines for action. The programs are prepared by the commission and are adopted by the European parliament and council.[15]

The EU's programs for the environment have the overall goal of sustainability, defined as the initiation of "changes in current trends and practices that are detrimental to the environment, so as to provide optimal conditions for socio-economic well-being and growth for the present and future generations."[16] Described below are the first six programs, with emphasis on the Fifth and Sixth Environmental Action Programs.

The First Environmental Action Programme (1973–1976)

The First Environment Action Program basically provided for remedial actions at the community level. The need to study the impact of human activities on the environment of the community was accepted as a necessary step toward understanding environmental influences. Several guidelines were adopted in the First Action Program including recognition of the importance of prevention and acceptance of the polluter-pays principle. A need to provide for public participation in environmental decision making and to inform and educate the public was also recognized.

The Second and Third Environmental Action Programs (1977–1981, 1982–1986)

The Second Action updated and expanded somewhat upon the first. The Third Action Programs began as the community was recognizing that a preventive approach was important to dealing with environmental problems.[17] The importance of environmental limits was becoming recognized, and areas of increasing concern, such as acid rain, came into focus. However, the implementation and enforcement of environmental legislation during this time was often poor, so little success was accomplished.[18]

The Fourth Environmental Action Program (1987–1992)

The Single European Act (1987), Article 130r, made environmental protection a community goal and formally extended the principle of subsidiarity to cover environmental policy. Article 100a (3) further provided that where health, safety, or the protection of consumers and the environment are concerned, the commission may base any decision on a high level of protection.[19] Thus, by the time of the Fourth Environmental Action Program, the need for environmental protection had become a central part of community policy making. The commission had become convinced that "the establishment of strict standards for environmental protection is no longer merely an option; it has become essential."[20] The text of the program points out that the

European Council had concluded (in its March 1985 meeting) that "environmental policy can contribute to improved economic growth and job creation" and were not simply added costs of production.[21]

The Fourth Environment Action Program placed priority on the implementation and enforcement of community legislation, focusing on substances and sources of pollution. It included a number of specific objectives in various segments of industry. In the area of agriculture objectives included making farmers more aware of environmental issues and maintaining the protection of the environment and landscape. Under regional policy, an adherence to environmental standards when planning economic development projects in disadvantaged areas was noted as being "of particular importance."

The Fifth Environmental Action Program (1993–2000)

Entitled "Towards Sustainability," the Fifth Action Program on the Environment covered the years 1993 through 2000. While the first four action programs provided lists of actions and principles, the Fifth Program provided a long-term strategy, as well as shorter-term objectives. The Fifth Action Program was the first of the environmental programs to directly address the issue of sustainability. "Towards Sustainability" describes the features of sustainability as:

- ➢ to maintain the overall quality of life;
- ➢ to maintain continuing access to natural resources;
- ➢ to avoid lasting environmental damage;
- ➢ to consider as sustainable a development which meets the needs of the present without compromising the ability of future generations to meet their own needs."[22]

To move toward sustainability, the Fifth Action Programme attempted to change practices that were viewed as leading to the depletion of natural resources and damage to the environment. To accomplish this, a broad range of instruments were directed at principal economic sectors, including energy, agriculture, industry, transportation, and tourism. In relation to the principal economic sectors, seven themes were addressed. These included climate change, acidification and air quality, the urban environment, coastal zones, waste management, management of water resources, and the protection of nature and bio-diversity. The plan was designed to sustain and benefit both the environment and the economic sectors. Figure 3.5 provides a list of the five target sectors and seven themes and targets of the Fifth Environmental Action Program.

The program focused on broadening the range of environmental policy tools from predominantly command-and-control mechanisms toward market-based instruments and voluntary approaches. Market-based instruments are designed to work within the structure of the market to produce the desired outcomes. These instruments include environmental taxes, subsidies, and transferable pollution permits. The market-based approach focuses on "getting the prices right," that is, adjusting subsidies and taxes to correctly price environmental resources in order to provide incentives (or disincentives) for the sustainable use of environmental resources.[23] Voluntary approaches include eco-labeling and eco-audits.

Eco-labels were established by the community (Council Regulation No. 880/92) to promote better consumer decisions by increasing information. The eco-label indicates an environmentally friendly product that does not violate EC regulations. The label may be applied for by either the manufacturer or the importer, who applies to the appropriate authority in the member state. The award is given to individual products based on a life cycle analysis of the environmental impact of the product group in which the product is placed.

Figure 3.5.
Fifth Environmental Action Program: Targets, Sectors, and Themes

Target Sectors

Industry
Energy Sector
Transport
Agriculture
Tourism

Themes

Climate Change
Acidification and Air Quality
Urban Environment
Coastal Zones
Waste Management
Management of Water Resources
Protection of Nature and Bio-Diversity

Source: Adapted from *"Towards Sustainability"*, *Official Journal of the European Communities*. No C 138/5 (175.93). Accessed 2 November 2004. Available from http://europa.cu.int/comm/environment/actionpr.htm.

In 1993, to further encourage firms to be responsive to environmental concerns, the commission adopted a voluntary environmental auditing system for industry (Council Regulation No. 1836/93). Its goal is to provide a standard for environmental management.[24] The eco-audit requires the company to incorporate environmental standards into their production processes. Member states coordinate the system and develop a list of firms who have been verified as complying with the regulations. Companies may choose whether to register and once registered, must follow certain rules. This system applies by site, not by company so that a company may register a number of its sites, but not others, as it desires. Once registered, a preliminary environmental review is conducted. The company bases its environmental protection system on the results of this review. Under the Eco-management and Auditing Scheme Regulation, each member state is required to set up a national system for implementing the eco-auditing scheme. The commission publishes a list of participating companies, the Registered Members List.[25]

An evaluation of the results of the Fifth Environment Action Program, entitled *Global Assessment,* was undertaken in 1996. The report was designed to examine progress in the principal sectors noted above. According to the report, progress was most evident in industries in which legislation had previously existed; progress was least evident in agriculture and tourism. In transport, progress was being made, but an increase in the number of vehicles was overwhelming the progress made in reducing vehicle emissions. The report concluded that "further efforts need to be undertaken in integrating environmental objectives into other policy areas and that stakeholders and citizens have to become more involved."[26]

As discussed in chapter 2, the public is offered opportunities to provide feedback to the EU regarding policies and their implementation. Accordingly, input was solicited from various sectoral groups, nongovernmental organizations, and other stakeholders regarding the Fifth Environmental Action Program.

The Sixth Environment Action Program (2001–2010)

The Sixth Environmental Action Program, entitled "Environment 2010: Our Future, Our Choice," takes a broad approach to the challenge of improving environmental policy. It identifies four priority areas: climate change, nature and biodiversity, environment and health, and natural resources and waste.[27] This program focuses on generating more input from citizens and more cooperation from business, on better implementation of existing legislation, and on integrating environmental interests into all relevant policy areas. Its goal is to "promote the integration of environmental concerns in all Community policies and contribute to the achievement of sustainable development throughout the current and future enlarged Community. The Programme furthermore provides for continuous efforts to achieve environmental objectives and targets already established by the Community."[28]

The Sixth Action Program lists several specific objectives to be pursued. One objective seeks to stabilize the atmospheric concentrations of greenhouse gases, with a long-term target reduction of 70 percent. Another seeks to protect and restore natural ecosystems, halt the loss of biodiversity, and protect soils against erosion and pollution. A third focuses on reducing the risks to human health due to environmental contaminants. Finally, the program calls on the union to use natural resources in a sustainable manner, that is, "to ensure that the consumption of renewable and non-renewable resources does not exceed the carrying capacity of the environment."[29]

The primary guiding principles of the program are to be based on the polluter pays principle, the precautionary principle and preventive action, and the principle of rectification at source. Thus the guiding principles laid down in the First Environmental Action Program continue as foundations for policy objectives in the Sixth Program. The Sixth Program also continues to promote better information for consumers by continuing to encourage eco-labeling and promotes a green public procurement policy, which allows environmental characteristics to be taken into consideration.[30]

The Sixth Environmental Action Program is only in its third year as of this writing, thus we cannot yet judge how successful it will be. However, it is important to note that there are significant challenges to be addressed. For instance, EU greenhouse gas emissions have risen for the second year in a row, moving the EU further from its target reduction.[31] This is one area in which the Environment Action Program will have to make serious inroads if the EU is to be successful in meeting its long-run objective of an 8 percent reduction in emissions by 2008–12. In order to better analyze the situation regarding air pollution, the Clean Air for Europe (CAFÉ) program was launched. Its goal is to "develop a long-term, strategic and integrated policy advice to protect against significant negative effects of air pollution on human health and the environment."[32] The first results of the CAFÉ program are expected to be available in early 2005. (We discuss greenhouse emissions in more detail in chapter 5.)

We can see from this overview of Environmental Action Programs One through Six progressively increasing attempts to address environmental problems. Policy follows society and during the period (from 1973 to the present) the directives of the EU—and the environmental action programs, have reflected a Europe that has increasingly (although to varying degrees in different countries) become aware of and concerned for environmental protection.

SUMMARY

Substantial change has taken place in the member states of the EU as environmental action has become part of the union's overall strategy. The European Union has taken major steps toward improving environmental quality. Policies have come to focus more explicitly on environmental concerns, yet the integration of environmental policy throughout the union is not complete. Implementation and enforcement activities are not uniformly applied throughout the union. Some nations have moved more quickly than others toward environmental improvements as legislated by the community. The European Union will need to focus on consolidating support among all member states if the goal of sustainability is to be achieved. The acceptance of central and eastern European countries into the EU will also have a substantial impact on environmental matters.

In the next chapter we examine some specific environmental issues in the EU, focusing on successes and failures in EU environmental policy and examining underlying causes for EU environmental success.

CHAPTER 4

SUCCESSES AND CHALLENGES IN EUROPEAN UNION ENVIRONMENTAL POLICY

INTRODUCTION

The focus and concerns of the EU have evolved over time; now the environment has taken an important place among other matters of interest to the community. Europe has seen improvement in many areas since the inauguration of the efforts of the EU toward environmental protection and sustainability. Many of the efforts directed toward environmental protection and improvement are in their early stages, and work continues in those areas, including the rebuilding of European forests and limiting the release of chemicals that lead to climate change. In other areas, such as the reduction of air pollution caused by automobile emissions, directives have been established for twenty years or more. Some of the policy efforts have been successful in generating environmental improvements; some have not. As we saw in the previous chapter, the EU has created the European Environmental Agency (EEA) in large part to evaluate the EU's environmental progress and provide the member states with reliable information. Table 4.1 provides a listing of several environmental areas that have been evaluated by the EEA. The table provides the EEA's estimates of how successful policy efforts have been and whether those policies have translated into environmental improvements.

In this chapter we will examine cases that illustrate environmental areas in which the EU has achieved varying levels of success, as well as areas in which the EU is still struggling to meet environmental challenges. We begin with the simplest case, that of an endangered species of bird, the Fea's petrel. Then we consider a protected area in Spain, the Riaza River Gorges, in which development is being promoted. We examine successful efforts at air pollution control in Finland and less successful efforts to control pesticide pollution of groundwater in Denmark. Finally, we analyze a more complicated situation, involving more than one country and an important economic sector, fisheries. We investigate the factors that are important in each case in meeting or failing to meet the challenges facing the environment.

Table 4.1:
The European Environment Agency's Evaluation of EU Environmental Policies, 1998

Environmental Area	*Policy Success*	*Environmental Success*
Climate change	Moderate	Little
Stratospheric ozone depletion	High	Little
Acidification	High	Moderate
Tropospheric ozone	Moderate	Little
Chemicals	Moderate	Moderate
Waste	Little	Little
Biodiversity	Moderate	Little
Inland waters	Moderate	Moderate
Marine and coastal environment	Moderate	Little
Soil degradation	Little	Little
Urban environment	Moderate	Moderate
Technological and natural hazards	High	High

Source: Adapted from European Environment Agency, *Europe's Environment: The Second Assessment* (Oxford: Elsevier Science, 1998), 16.

CASE STUDIES

In this section we will attempt to evaluate the success of the EU in environmental policy by examining several cases where the community has acted to protect the environment. First, we will discuss the relevant EU and national regulations that apply in each case examined. Then we will analyze the physical, institutional, and economic forces that come into play in the case. In particular, we will consider the role of economic forces that motivate individuals and nations to move toward certain goals and away from others. Individuals, of course, are generally assumed to be motivated by self-interest. Historically, the motivating force has been ascribed to economic factors, that is, income. However, in recent years, other factors such as altruism and satisfaction from enjoying nature have been viewed as other potential motivators. Thus it is possible for individuals to move toward sustainable farming methods, for example, in spite of a drop in income that may be associated with those methods. However, we can also assume that there are most likely limits to the income that individuals are willing to trade off in exchange for environmental benefits. This may be particularly true in the case of benefits that are not limited to the particular individual who suffers income loss to produce them but can also be consumed by free riders, that is, those individuals who are able to consume a good or service without paying for it.

After reviewing the specific national and EU regulations associated with each case, we will examine the economic incentives or disincentives relevant to the actors associated with the case. We will find that, as we suggested in chapter 1, the structure of the regime has influenced the actors in certain ways. Finally, we will evaluate the implications these economic factors have regarding the success of the EU policy.

Portugal: The Fea's Petrel

The first case we examine is that of the Fea's petrel. The Fea's petrel (*Pterodroma feae*) is a very rare, threatened bird, which exists only on Bugio, the southernmost of Portugal's Desertas Islands and on some of the Cape Verde Islands. The Fea's petrel was originally identified as a species of *Pterodroma mollis* in 1853, and as a distinct species, *Pterodroma feae*, in 1900. The total world population of Fea's petrel is currently estimated to be one thousand breeding pairs; approximately 20 percent of these are located in the Desertas Islands.[1]

The eating habits of the Fea's petrel are not well documented, but it is thought that their diet consists mainly of fish and crustaceans. Breeding occurs where the earth is thickly covered with grass; the birds burrow about thirty to sixty centimeters below the surface to nest. The birds that inhabit Portugal's Desertas Islands are in danger due to overgrazing of their habitat by goats. Besides loss of vegetative cover, overgrazing leads to soil erosion, reducing the locations with sufficient soil depth available for the birds to nest. Rabbits are also a concern as they sometimes invade and modify the birds' borrows. This causes stress to breeding pairs, who may abandon their nests as a result of the rabbits' intrusion. The suggested solution to this problem is to relocate the rabbits and goats to other islands in order to provide a safe habitat for the petrel. The birds' only natural enemies in the Desertas are the yellow-legged gulls, who feed on the petrel.

The Fea's petrel is listed in the EU's *Wild Birds Directive Annex I*, which lists the 182 most endangered bird species in the EU. EU Directive 79/409/EEC on the conservation of wild birds and its amendments establishes a system of protection for all naturally occurring wild bird species in the territory of the member states. The purpose of the directive is to protect and manage the species and to regulate any hunting and capture of the birds. Protection of the birds' nests, eggs, and habitats is included under the directive. Article 3 specifically calls for the preservation or improvement of bird habitats as necessary for achieving the goals of the directive. Article 4.1 of the directive states that "The species mentioned in Annex I shall be the subject of special conservation measures concerning their habitat in order to ensure their survival and reproduction in their area of distribution."[2] Beyond those measures, amendment Directive 92/43/EEC provides that, if necessary, public interests, "including those of a social or economic nature" may be overridden and "the Member State shall take all compensatory measures necessary to ensure that the overall coherence of NATURA 2000 is protected."[3]

The NATURA 2000 Network is a list of sites that contain habitat types and species of importance to the community. NATURA 2000 is based on two directives: Council Directive 79/409/EEC, discussed above, and Council Directive 92/43/EEC on the Conservation of Natural Habitats and of Fauna and Flora (the Habitats Directive). As we will recall, directives are binding to the member states to which they are addressed but allow the member states to determine the method of implementing the directive. The Habitats Directive covers 253 habitat types, 200 animal species, and 434 plant species.[4] Under the directive, each member state was required to perform a national assessment of the habitat types and species in that country and submit the list to the commission. Together, the two directives comprising NATURA 2000 provide the framework for the protection of wildlife and their habitats in the EU. The network is made up of special protection areas, designed to protect both migrating birds and the bird species included in Annex I, and special areas of conservation, set up to protect the plant and animal species and habitat types listed under the Habitats Directive. By 1995, there were over one thousand special protection areas, covering an area the size of Belgium and the Netherlands combined, which amounts to about 3 percent of the total EU land area.

The Portuguese Red Data Book, a national listing of the conservation status of plant and animal species, classifies the Fea's Petrel as vulnerable; thus it is a protected species under Portugal's Decreto-Lei 75/91. (National Red Data Lists are generally based on the criteria developed by the World Conservation Union (IUNC), a group of states, government agencies, nongovernmental organizations, and scientists.)[5] The Portuguese government has designated the Desertas Islands as a special protection area under Article 4 of the EU Wild Birds Directive.[6] The EU has provided funding for the Desertas Islands Nature Reserve since its creation in 1990 through Actions by the Community for Nature (ACNAT). ACNAT was established by Council Regulation 3907/91; it was later superceded by the Financial Instrument for the Environment (LIFE), which began in 1992 under Council Regulation 1973/92 and was later revised under Regulation 1655/2000.[7] LIFE finances projects including nature conservation, development of new methods for the protection of the environment, and actions for promoting sustainable development in third countries.[8] Over two thousand projects were funded through LIFE in its first ten years.[9]

Protection measures for the Fea's petrel have been quite successful. In the mid-1980s there were an estimated 75 pair of Fea's Petrel in the Desertas; today there are estimated to be 150 to 200 pair.[10]

As suggested earlier, we have started with a simple case of environmental protection in the EU. The Fea's Petrel is a noncommercial bird, living in islands that are not viewed as economically valuable to humans. Thus, other uses of the Desertas do not compete with their use as a habitat for the Fea's petrel. In addition, the EU has provided funding for the nature reserve, alleviating the need for funding from Portugal or other sources. In this case, setting aside the nature reserve under EU directives did not face many obstacles. The case of the Fea's Petrel illustrates a relatively simple EU environmental success. The directives and institutions involved acted as intended. But then in such a situation, with no competing economic or other interests, one would expect success in this case. As we shall see in the other cases described, decisions regarding uses of environmental resources are rarely so simple. The next case, that of the Riaza River Gorges area of Spain, illustrates the situation in another protected area. In contrast to the Desertas Islands, however, development of the Gorges area for human economic benefit is actively being encouraged by the EU.

Spain: The Riaza River Gorges

The Riaza River Gorges, located in the northeastern administrative region of Segovia, has been a protected area since 1989. The area is the home of a variety of endangered habitats, animal, and plant species, such as Pyrenean muskrats, Egyptian vultures, and sabina, an extremely rare plant species.[11] The area is sparsely populated, having fewer than four hundred inhabitants, most of whom are elderly.

Under Natura 2000, the Gorges area has been designated as economically disadvantaged. In 1999, Spain's gross domestic product per capita of US $17,300 was only three quarters of that of the leading European economies; the Spanish unemployment rate averaged 16 percent.[12] It is hoped that economic development in the Riaza River Gorges locale can be realized by building a tourism industry on the basis of the protected area's appeal to nature enthusiasts.

Major funding for economic development in the Gorges area was provided for by the European Social Fund (ESF). ESF is one of the EU structural funds, which are designed to help regions within EU member states that are relatively underdeveloped economically.[13] ESF funding focuses on programs that develop employment skills.[14] As part of the effort to develop the Gorges area's economy, the European Social Fund provided more than 161,000 Euros (US $165,830)[15] to fund programs impacting over thirteen hundred people, from both within and outside the Gorges area.[16] Funds were used for several purposes including increasing awareness among the local residents, providing environmental education for the area's children, and training volunteers in the Gorges refuge area. By 2002, over five hundred volunteers had participated in various activities at the refuge. The development of existing hotels and tourist destinations, such as museums and visitor centers, will further stimulate the area's economy. The manufacture of locally made goods and crafts is also being encouraged. Along with local farm production, these activities are expected to help provide a solid foundation for the area's economy. As this project is relatively new, it is too early to determine the long-term impact, but thus far results are favorable.

The framework for sustainable rural development in the community is provided for by Council Regulation 1257/1999 on Support for Rural Development from the European Agricultural Guidance and Guarantee Fund.[17] This regulation includes such measures as a less-favored areas scheme (for economically disadvantaged areas), funding for the training of young farmers, assistance for conversion of agriculture, and investment aid for marketing facilities. For example, support is available to farmers who use sustainable production methods for at least five years. The aid is calculated based on foregone income and includes an incentive to encourage farmers to make environmentally sustainable farming decisions. Funding is limited, however, to a maximum of 900 euros (US $927) per year and varies according to the crop. Farmers in less-favored areas, such as the Riaza River Gorges area, can receive funds to encourage their continued use of sustainable methods. Farmers are also eligible for funds designed to replace income losses associated with the implementation of EU environmental regulations. Additional financing is available for the development of rural areas, including the promotion of tourism and craft industries.[18] The Riaza River Gorges area and its residents thus have financial resources available to help in development of the local economy.

This case illustrates a situation in which the economic incentive supports protection of the region's wildlife habitats. Residents of the Gorges area can benefit economically from the protected area, through the development of tourism and related industries. Thus there is an incentive to protect the source of their income, that is, the protected area. In addition, local farmers can receive funding to help implement sustainable production methods and can benefit from the local sale of their products in the revitalized economy of the area. In this way, the EU is using positive economic incentives to encourage sustainability and the preservation of environmental resources.

Our next case examines another area associated with farm production, the case of groundwater pollution in Denmark. As we will see, this case becomes more complicated because so much of Denmark is under cultivation; thus it is an important segment in the nation's economy.

Denmark: Pesticides

Pesticides are economically important because of the increased agricultural yields that can result from their use; however, they can also be harmful to human health and the environment. Most of Denmark's soil is podzolic and infused with acid solutions that drain its minerals. Thus, heavy fertilization is required for successful agricultural production. The use of pesticides in Denmark is significant because approximately 54 percent of the land area of Denmark is cultivated.[19] In 1996, 3,750 tons of pesticides were used in the country.[20]

While pesticides have increased agricultural yields in Denmark, there have been environmental costs. The quantity of pesticides in Denmark's groundwater is quite high. Of 307 sites monitored in the 1990s, 25 percent showed increases of pesticide levels from the early 1990s to the late 1990s, 61 percent showed no change, and only 13 percent showed declining pesticide levels.[21] Eighty-four samples taken from 1994 through 1995 exceeded the amount of allowed pesticides in 35 percent of the samples. While these sites exceed the allowed limits, other areas may also exhibit lower concentrations of pesticides.[22] These findings are of considerable concern as over 75 percent of the water for public use is supplied from groundwater sources.[23] The pesticides most commonly found in the groundwater are phosphorus and nitrogen from farming; pesticide levels are generally the highest during spring and autumn, when pesticides are applied to farm fields.

The EU has passed legislation regarding the use of chemicals, including pesticides.[24] EU regulations divide pesticides into two groups, plant protection products and biocidal products. Both groups of products must be assessed and authorized before they can be marketed. Because some pesticides are released into the environment, particularly water, they are also covered under water legislation. Both EU chemical and water regulations related to pesticide pollution of groundwater are discussed below.

Chemicals are governed under three directives and include both existing and new substances. Provision of data is covered under Directive 67/548, which encompasses new substances, and Regulation 793/93, for existing substances. Under this directive, manufacturers or importers must provide data on new substances before they can be marketed; substances already on the market can continue to be sold while the data is being provided. Directive 93/67 and Regulation 1488/94 cover risk-assessment procedures related to chemicals. Risk assessment is done under the auspices of the Directorate-General of the Environment and provides the basis for marketing or use restrictions. Directive 76/769 limits marketing and the use of dangerous substances.[25]

Community water policy is directed toward the protection of water used for drinking and other human uses, for the protection and preservation of the aquatic environment, and for the reduction of natural disasters associated with water, such as drought and floods.[26]

Directive 2000/60/EC provides the framework for EU water policy.[27] The directive covers the protection of waters, including groundwater, surface waters, and coastal waters. Its aim is to prevent and reduce pollution, promote sustainable use of water, and protect the aquatic environment. The framework directive for water policy includes measures designed to internalize the external costs associated with water usage, for example, requiring "that the various economic sectors contribute to the recovery of the costs of water services, including those relating to the environment and resources."[28] Member states are also to provide incentives designed to ensure that water resources are used efficiently. However, direct limits or quotas on the emission of certain substances into water will be imposed in order to eliminate or reduce those substances that present significant risks to the aquatic environment. In 2003, the commission is expected to publish proposals for specific measures to control groundwater pollution.

Farms are the main source of pollution of water by nitrates in Denmark (and throughout the EU). The environmental effects of the use of chemical fertilizers are targeted by Council Directive 91/676/EEC concerning the protection of waters against pollution caused by nitrates from agricultural sources. (Other directives address the pollution of groundwater from other sources.) Under this directive, member states are required to monitor both surface and groundwaters which are, or could be, affected by pollution, and to identify vulnerable zones. The states must establish voluntary codes of good agricultural practice, such as limiting the use of nitrogen-containing fertilizers and livestock effluent. (The European Environmental Bureau, a federation of nongovernmental organizations that has consultative status with the Council of Europe, the European Commission, Parliament, and the Economic and Social Committee of the EU, as well as OECD and the UN Commission on Sustainable Development, has recommended that the EU institute mandatory reduction plans for all member states, including a target reduction of 50 percent within ten years from the base year.[29]) Member states are authorized to take additional measures as needed to attain the objectives of the nitrate pollution directive.[30] A commission report on the implementation of the directive indicates many of the member states failed to meet their obligations to transpose the EU requirements into national plans by the December 1993 deadline. However, Denmark was one of the states that did set up action programs at the national level.[31] In addition to coverage under the directives noted, the sustainable use of pesticides has also been included as an issue of major importance under the Sixth Environment Action Program.

EU rural development policies include positive economic measures that provide payments for commitments by farmers to use more sustainable methods. Organic farming, for instance, avoids the use of pesticides. As we noted above, the maximum amount available to farmers to encourage the use of sustainable methods is only US $927. However, the median income for Denmark's farmers is about 200,000 Danish kronor (US $28,000); thus, the amount available to offset income losses may not be sufficient to encourage Danish farmers to sacrifice income in order to reduce pesticide applications.[32] As we suggested above, farmers are likely to evaluate the potential lost income against the funds available from EU sources to offset those losses. In addition, farmers may be willing to give up some income in order to pursue sustainable farming methods; however, the extent of lost income they are willing to trade off will be limited. Thus, the positive economic incentives provided for by the EU legislation may not be sufficient to elicit the desired responses on the part of farmers, particularly higher income farmers such as those in Denmark. Additional positive incentives may be required; alternatively, negative incentives, such as increased taxes on pesticides, may be necessary to reduce pesticide use.

It is still too early to evaluate the full extent of the success (or ultimate failure) of EU environmental initiatives on the regulation and control of pesticides in Denmark. Nonetheless, we have to classify this case study as an example of an early success of EU environmental policy making. Largely in response to EU directives, Denmark developed and implemented a nationwide plan for the containment of pesticides. In addition, EU programs provide the framework and have started to provide the payments necessary to move farmers away from pesticide-intensive agriculture. The EU laid the foundation for important environmental reform in this case. The ultimate success of pesticide reduction in Denmark will now depend on future funding and implementation of existing programs.

Our next case extends the range of actors involved in the process to encompass almost the entire population of Finland. In this case, we examine Finland's efforts to control pollution from vehicles. In such a situation, the number of actors becomes much larger due to the many sectors of the economy that use automobiles, either for personal use, commuting to work, or hauling materials and merchandise across Finland's roadways.

Finland: Air Pollution from Gasoline-Powered Engines

In 2000, Finland had 77,895 kilometers of roadways[33] and 413 passenger cars per one thousand inhabitants (up from 155 in 1970).[34] This increase in the number of automobiles occurred in spite of an 89 percent tax added to the price of a new passenger car. Ninety percent of automobiles in Finland are gasoline powered.

Finland's environmental legislation, reducing pollution caused by gasoline-powered engines, follows relevant EU legislation. A 1995 amendment to the Finnish Constitution (section 14a) includes reference to environmental responsibility. "Responsibility for nature and its biodiversity, for the environment and for our cultural heritage is shared by all. Public authorities shall strive to ensure for everyone the right to a healthy environment as well as the opportunity to influence decision-making concerning the living environment."[35] Finland uses positive economic incentives in the form of tax refunds for the use of renewable energy sources, such as wind and wood-based fuels. Small-scale hydropower and peat-based power production are also eligible for refunds. Negative incentives, such as taxes, are also used to encourage the desired environmental results.

Finland has applied strong negative incentives to control pollution associated with the use of automobiles to focus on reducing the emission of lead from gasoline-powered engines. Heavy taxes were applied to leaded gasoline through the 1980s and 1990s to encourage users to switch to vehicles using unleaded fuel. Taxes on unleaded gasoline rose 60 percent. (Taxes on diesel oil increased 44 percent, and taxes on other fuels rose 75 to 95 percent during the same period).[36] Excise taxes on motor fuels were almost ten times higher than taxes on other fuels and rose steadily during the 1990s. Excise taxes do not include CO_2 taxes. The CO_2 tax, introduced in 1990, is an additional environmentally based tax determined on the basis of the fuel's carbon content. (Transport accounts for 24 percent of total CO_2 emissions from fossil fuels in the EU.)[37] In 1989, positive tax incentives were provided for vehicles using catalytic converters.

Finland provides an interesting case because its national direction regarding automobile emissions mirrored that of the EU. Finland's original environmental legislation dates from the 1970s and 1980s. In 2000, Finland passed a new Environmental Protection Act, revising their pollution control legislation to harmonize with the EU's relevant legislation. (As noted above, in 1995 the Finnish constitution had been amended to include environmental goals.)

Finland's legislation is based on EU directives. More than fifteen different EU directives, dating from 1974, focus on automobile emissions that pollute the air.[38] Limits of various emissions and requirements for newly manufactured vehicles are provided in the directives. The EU's Sixth Environment Action Program focuses on environment and health, including air pollution, as one of its main target areas.[39] The community is focusing on the reduction of CO_2 emissions from passenger cars in order to meets its goals of reducing new passenger car emissions to 120 grams CO_2 per kilometer by 2010. (As a party to the United Nations Framework Convention on Climate Change, the EU agreed to stabilize its emissions of CO_2 at 1990 levels by the year 2000.)[40] As of 2002, new cars marketed in the union were producing 140 grams CO_2 per kilometer.[41] In order to achieve the objective to reduce CO_2 emissions by an additional 20 grams per kilometer, three steps have been taken. Agreements have been made with automobile manufacturers to use improved technologies to reduce CO_2 emissions, market-oriented measures are being introduced to influence

Table 4.2.
EU Programs for Integrating Environmental Concerns into the Energy Sector

➢ ETAP (Environmental Technology Action Plan)

- Surveys technologies that can help to address environmental problems
- Monitors market developments and energy trends to discover barriers hindering the development and implementation of specific technologies[44]

➢ SYNERGIE

- Focuses on the strengthening of international cooperation in energy matters

➢ ALTENER

- Encourages the use of renewable energy sources

➢ SAVE and SAVE II

- Encourage the rational and efficient use of energy resources
- Aim at the stabilization of CO_2 emissions and the attainment of improvements in energy efficiency

➢ JOULE-THERMIE

- Demonstrates and promotes new, clean, and efficient energy technologies

Source: European Commission, "Framework Program for Actions in the Energy Sector (1998–2002)." Accessed 6 June 2003. Available from http://europa.eu.int/5cadplus/leg/-en/lub/127024.htm; European Commission, "Energy: The Integration of Environmental Considerations into the Energy Sector." Accessed 6 June 2003. Available from http://europa.eu.int/comm/environment/environment/-act5/chapt1–2.htm.

consumer choice, and efforts are being made to increase consumer knowledge of automobile fuel economy.

Finland's use of heavy taxes to reduce pollution from gasoline-powered engines by targeting the reduction of automobile use has been successful. This case illustrates the potential for negative incentives to be effective in eliciting the desired response. Increased prices on fuels encourage individuals to drive less, to purchase more fuel-efficient automobiles, and/or to use public or other forms of alternative transportation. The high tax on the purchase of new automobiles, however, may not be as effective if we assume that new automobiles are more fuel efficient and less polluting than the older models. The tax on new autos may lead people to purchase older, less environmentally desirable models.

Our next case focuses on the move to develop and use more renewable energy sources. We examine the use of wind energy in Ireland.

Ireland: Wind Energy

Our second case involving the energy sector focuses on the development of renewable energy in Ireland. Production of energy from renewable sources is one of the key areas currently being addressed in the EU in order to reduce energy consumption and associated CO_2 emissions, which are detrimental to the environment.[42] The EU has created several programs that focus on the integration of environmental concerns into the energy sector; table 4.2 provides a short description of selected programs. For example, one of the EU's early integrated programs, the Framework Program for Actions in the Energy Sector (1998–2002), focused on three main areas; the security of energy supplies, competitiveness, and environmental protection. Implementation of the program was based on six elements, including the promotion of renewable energy sources through the Altener program. Along with activities undertaken at the EU level, member states were expected to focus on increasing energy efficiency. Specifically, member states were to plan for greater exploitation of renewable energy sources and to move toward increasing the use of energy from renewable sources, compared to fossil fuels. This focus has taken many forms among the member states, for example, in Austria, the development of hydropower and solar energy; in France and Finland, the use of biofuels; and in Denmark, forest residues.[43]

In Ireland, the expansion of renewable energy sources took the form of the development of wind energy. Wind energy is a form of renewable energy that is readily available in Ireland but was not being used to any significant extent. An increase in the use of energy from renewable sources can be particularly important to a country such as Ireland, which uses large quantities of imported fuels. (Table 4.3 provides a summary of the use of various forms of energy in Ireland in 1998.)

In response to the EU impetus toward the use of renewable energy, Ireland's National Development Plan (1994–1999) identified as one of its goals, the "production of as much of the country's energy requirement from indigenous sources as is economically possible."[45] This theme builds on the objective under the EU's Fifth Environmental Action Program, that is, the tripling of the production of energy from renewable sources, and the EU program for actions in the energy sector, described above. As part of its development plan, Ireland set a target of increasing electricity generation from renewable sources. In order to achieve the target, Ireland's Electricity Supply Board initiated a scheme to develop alternative energy sources, including hydro, wind, landfill gas, biomass, and waste, by the end of 1997. Ten wind power projects were approved for development. The original target of an

Table 4.3.
Energy Sources in Ireland, 1998

Energy Source	Percent of Total Energy Provided by Source
Coal	16.2
Peat	7.7
Oil	55.6
Natural Gas	18.5
Renewables	2.1

Note: Percentages do not add to 100 due to rounding.

Source: Adapted from Sustainable Energy Ireland, "Ireland's Total Energy Requirement 1998." Accessed 27 May 2003. Available from http://www.environ.ie/search/search-index.html.

increase of fifty-five megawatts of electricity from sustainable sources was achieved; the target for the period 2000–2005 is an increase of five hundred megawatts from sustainable energy sources. Most of Ireland's future renewable energy is expected to come from wind power.

One of the representative wind projects in Ireland was the Kilronan wind farm, developed by the Irish company Kilronan Wind Farm Limited. The farm is located in County Roscommon, near existing access roads and power lines. Built in 1997 and consisting of ten turbines, by 1998 it was generating fourteen gigiwatts (1 billion watts) of electricity and was considered a success.[46] As a result of the focus on wind energy installations in Ireland, including the Kilronan installation, a twelve-fold increase in the generation of wind energy occurred between 1993 and 1996.

To further focus on the development of renewable energy, the Irish Energy Center set up its Renewable Energy Information Office in 1995, with the goal of promoting the use of renewable energy. In addition, as a part of its planning mission, the Irish Ministry of the Environment's Planning Section has published a reference manual for the development of wind energy.[47] While large-scale funding for the development of renewable energy is not available from the government of Ireland, partial support in the form of tax relief is available.

As noted earlier, the EU has set a goal of increasing the use of power from renewable sources and decreased reliance on imported energy. According to the multiannual program for action in the field of energy, which covers the years 2003 to 2006, "the European Union is becoming increasingly dependent on energy from non-EU countries. . . . The Union therefore wishes to reduce its dependence and improve its security of supply by promoting other energy sources and cutting demand for energy. Consequently, it is putting the accent, above all, on improving energy efficiency and promoting renewable energy sources."[48] The program focuses on four areas: rational use of energy in building and industry, including demand management; the development of new and renewable energy sources; energy efficiency and the diversification of fuels used in transport, and the promotion, at the international level, of renewable energy sources and energy efficiency in the developing countries.[49] The EU Commission has proposed a budget of 215 million euros for implementation of the program.

While funding from the governments of member states is not always available, EU structural funds can be applied for to assist in the development of renewable energy sources; funding from the ALTENER program may also be available. The ALTENER program (Decision No. 646/2000/EC of the European parliament and of the council of 28 February 2000 adopting a multiannual program for the promotion of renewable energy sources in the community) was designed to promote the use of renewable energy sources in the EU.[50] Seventy-seven million euros were provided for the program to support various activities associated with the development of renewable energy.[51] The "First Review Report 2001" on progress made under the program indicated success in the priority areas, one of which was the promotion of electricity from renewable energy sources.[52] Activities under the Alterner program include studying the implementation of renewable energy sources, monitoring the development of renewable sources, creating or extending structures for the development of renewable sources, and implementing actions to facilitate the market penetration of renewable energy sources. Such projects are eligible for 50 to 100 percent financing by the community, depending on the type of activity. By 2010, at least 12 percent of energy in the EU is to be from renewable sources; currently renewable sources provide 15.3 percent of total electricity consumption in the EU.[53]

The economic incentive in this situation is the funding available for development purposes. In addition, there is a political incentive to become less dependent on imported oil. As we can see, member states, including Ireland, have implemented EU directives by developing different forms of renewable energy, based on the particular characteristics of the country. In Ireland, wind energy is abundant; thus, the Irish government has chosen to focus on developing this form of energy. As we have seen, the use of wind power has increased substantially.

North Sea Cod Fisheries: A Commons Problem

Numerous complaints have been brought against the EU concerning its Common Fisheries Policy. Seas at Risk, an independent nongovernmental coalition of European national and international environmental organizations concerned with the protection and restoration of the marine environment, has criticized the Common Fisheries Policy for failing to protect fish stocks.[54] Seas at Risk argues that the "available fishing capacity of Community fleets far exceeds that required to harvest fish in a sustainable way."[55] Some stock, such as cod, hake, and whiting is far beyond safe biological limits. Stocks of these species have declined 90 percent from the early 1970s to the late 1990s.[56] We will consider the case of cod in the North Sea as representative of the problems faced in fisheries policy and the ability of the EU to address that policy.

Evidence indicates a serious sustainability problem in cod fishing areas. A 1996 through 1998 study commissioned by the European Commission showed the total biomass of cod on the Flemish Cap to be at very low levels. Spawning stock was also found to be at a very low level. The study concluded that "the premature and continued exploitation until exhaustion of the most abundant cohorts of the last two decades by the trawl fleets... lead to lowest level for stock biomass on record."[57] Catches of cod in the North Sea, used by both EU and Norwegian fishing vessels, are only about one-sixth of what they were in the 1980s. In the 1980s, catches were about three hundred thousand tons per year; in 2000, they amounted to only fifty thousand tons.[58]

Generally, cod are caught by trawlers. The problem is that trawlers largely use small mesh nets. These small nets catch young fish, leaving fewer fish available for breeding stock. The last good year for spawning was reportedly in 1996, but many of those fish have been caught before reaching breeding age. In 2001, as a result of the findings noted above, the commission instituted emergency measures to protect the stocks that were considered in danger of immediate collapse. The control measures that were implemented focused on the gear used and were designed to reduce the catch of young fish in order to reestablish breeding stocks.[59]

The European Commission's Green Paper on the Future of the Common Fisheries Policy, released in 2001, acknowledged a need for improvement in the policy in order to restore a balance between fish harvests and protection of fish stocks.[60] In particular, black markets for fish have hampered efforts to protect the stocks. Often, when trawlers had already landed their quota of fish, additional fish caught would be sold on the black market. Annual quotas were thus circumvented, and EU policies became ineffective for protecting the resource.

The Common Fisheries Policy (CFP), based on Article 33 of the treaty establishing the European Community, covers all fishing, the farming of aquatic resources, and the processing and marketing of fishery products. The CFP also covers relations and agreements with nonmember countries and international organizations. The European commissioner of agriculture, rural development and fisheries oversees a directorate-general who is responsible for the Common Fisheries Policy and supervises the community's fisheries research program.

The EU's Council of Fisheries Ministers (composed of the fisheries ministers of the member states) sets total allowable catches (TACs) annually. The TACs are based on the recommendations of the European Commission and consultations with relevant noncommunity countries, such as Norway. This is the final step of a lengthy process. Initial data comes from the International Council for the Exploration of the Sea (ICES), an intergovernmental marine science organization. ICES collects data to assess the main fishery stocks targeted by commercial fishermen. The data collected is examined by the ICES Advisory Committee on Fishery Management, which is made up of representatives from EU member states and presented to the EU's advisory group, the Advisory Committee for Fisheries and Aquaculture (ACFA).[61] ACFA is made up of representatives of fishery and related stakeholders and covers the areas of resource management, aquaculture, markets, and trade. ACFA examines the assessment presented by the ICES Advisory Committee and presents the findings to the commission. The commission examines the data provided by ACFA, as well as reports by the commission's own Scientific Technical and Economic Committee on Fisheries, and makes recommendations to the Council of Fisheries Ministers, who then set the final TACs for the year.

Member states are responsible for ensuring that all vessels carrying their flags comply with the CFP regulations both within and outside of EU waters. Currently, the member state controls fisheries management within either a six- or twelve-mile zone, depending on the state. Within that zone, the member state may apply stricter conservation measures than community standards; however, those stricter measures will apply only to ships of the member state.

Negotiations on the rules of the common fisheries policy occur every ten years. The EU can recommend a ban on cod fishing in the North and Irish Seas, which will necessarily force a similar ban on other fish, such as haddock and whiting, because cod are regularly caught along with the targeted fish. There has been much dissention among the member states regarding proposals for changes in the CFP, particularly regarding management plans and reductions in capacity and fishing effort.[62]

Historically, politicians have been lobbied to keep quotas higher, and high catches were viewed as desirable, the stock of fish treated as inexhaustible, so such changes in policy are heavily debated, particularly as fisheries are such economically important areas in many countries. The proposed changes to the Common Fisheries Policy, for example, are estimated to cost about ten thousand to fifteen thousand jobs in Scotland.[63]

There is, however, funding available to help those harmed by changes in the CFP. The EU has set aside $600 million in aid to be directed to those who are economically harmed by fishing regulations.[64] Additional funds are available through the Financial Instrument for Fisheries Guidance, which is designed to "contribute to achieving a lasting balance between fisheries resources and their exploitation... [to] help revitalize areas dependent on fisheries and aquaculture."[65] This fund also provides monetary assistance for fishing vessel owners for scrapping vessels, permanently transferring (to a third country), and/or permanently reassigning vessels for other purposes, as well as providing financial help for fishermen for early retirement or training for other careers. The member state governments determine the allocation of available funding.

The case of fisheries illustrates the more complicated situation in which humans depend on the resource in question for their livelihood. Fisheries exemplify the characteristics of common pool resources. Fish stocks are mobile; they exist in waters that are not owned by any particular country but are available to all. The economic motivation is for the fisherman to catch as much fish as he or she can in order to secure the highest income for himself/herself. This is, however, true for all fishermen fishing the waters. No one country owns the waters or the fisheries stocks. As a result, no one is motivated to protect the resource for its long-term viability. As we have seen in our discussion of EU cod fisheries, this is exactly what has happened in the North Sea. In spite of TACs and other regulations, EU fisheries are in a state of crisis. In particular, this case illustrates the problem that sometimes arises with strict imposed limits: people will find ways to circumvent limits if they find it economically beneficial to do so. Therefore, such limits must have sufficient penalties, which must be imposed, if they are to be successful.

IMPLEMENTATION OF ENVIRONMENTAL REGULATIONS

While the EU has been successful in making environmental concerns an integral part of its policies, there are still problems of time lags associated with implementing EU policies. The Eighteenth Annual Report on Monitoring the Application of Community Law (2000) provides data on implementation problems in the environmental sectors.[66] There have been a large number of cases brought before the commission involving environmental matters. Of the two hundred cases brought before the commission in 2000, thirty-nine went before the Court of Justice; the Court of Justice delivered opinions or supplementary reasoned opinions in an additional 122 cases.[67]

The commission is responsible for ensuring that member states transpose community provisions in matching national provisions. The commission is also responsible for ensuring that environmental directives and regulations are properly applied in the member states. Of 543 complaints received by the commission regarding application of community environmental law in 2000, 30 percent were related to nature conservation, 25 percent to environmental impacts, waste concerns represented about 15 percent, water, 10 percent; other sectors accounted for the remaining complaints.[68]

Under community law, EU member states are required to transpose EU measures into national measures. When member states do not take the required steps and notify the EU of having met their obligations, the commission may start proceedings for member states' failure to notify the commission. The commission is also responsible for ensuring that environmental directives and regulations are properly applied.[69] Many of the complaints the commission received from individuals and nongovernmental organizations are petitions concerning incorrect application of EU law.

One particularly notable case of a member state's failure to meet EU environmental requirements involves Greece. In 2000, Greece was the first member state to be required by the European Court of Justice to pay a penalty for failure to act as required under EU law. The penalty payments stem from a 1992 decision by the Court of Justice, in *Commission v Greece*, C-45-91, in which Greece was determined to have failed to meet the requirements of Directives 75/442/EEC on waste and 78/319/EEC on toxic and dangerous waste to "ensure that solid, dangerous and toxic waste is disposed of without endangering human health and without harming the environment."[70] Article 10 of the EC treaty requires member states to cooperate in good faith with community decisions. Due to the country's continued failure to meet the requirements of the directive, the penalty payment of approximately US $19,000 (16,207 euros) a day, which is charged until the directives are met, was enacted.[71]

The Greek case involved a landfill in Crete, in which unregulated rubbish (including military and medical waste) had been dumped into a ravine close to the mouth of the Kouroupitos River. The dump burned continuously for over ten years due to the high organic content of the waste, while waste seeped into the nearby waters of the Mediterranean Sea.[72] The commission first acted on the situation after receiving complaints in 1987. The commission found that Greece had not developed waste disposal plans as required under EU directives 75/442 (Art. 4) and 78/319 (Art. 5).[73] In spite of a 1992 decision by the European Court of Justice, in which Greece was found to be out of compliance, Greece ignored continued demands to clean up the area.

There are a number of other cases in which environmental legislation has not been transposed into national law or correctly applied at the national level. For example, Directive 90/313/EEC on freedom of access to information on the environment is considered to be a very important piece of legislation in that it gives the public access to information so that citizens become contributors to environmental protection. Several member states have been found in violation of this directive. For example, in Case C-217/97, the commission issued a reasoned opinion against Germany based on Article 228 of the treaty for failing to comply with Directive 90/313/EEC. Court proceedings were also started against Austria for not completely transposing the freedom of access directive.

The most common complaint received by the commission in relation to the freedom of information directive is that of slow response on the part of national authorities, broad interpretation of the exceptions to the principle of disclosure and high charges for the information provided. The commission generally will direct individuals back to their national government, unless the case indicates a general transposition or application problem in the state.

In some states, the EU directive may not have been properly transposed. For example (case C-392/96), the court found that Ireland had not properly transposed the EU directive regarding environmental impact assessments. Ireland has set thresholds for certain types of projects; these were sometimes so high that projects with extensive environmental impact were not covered by the assessment procedure. When Ireland did not take the necessary measures (as required by the court's judgment), the commission sent the state a letter of formal notice under Article 228 of the treaty.

Numerous infringements in the area of air and water quality have been brought before the court, as well as cases involving the implementation of nature protection directives, noise, and chemical directives. For example, radiation protection measures are based on chapter 3, "Health and Safety," of the Euratom Treaty. This legislation was revised and made more comprehensive under Directive 96/29/Euratom on Basic Safety Standards. However, the commission found that only two member states (Finland and Italy) had provided a complete set of national transposing measures by the deadline set under the directive. Therefore, by summer 2000, the commission opened infringement cases against the remaining member states.[74]

Fisheries

Of particular note are the various infringements found in fisheries legislation. Numerous cases of fish catches over quotas have been found. The commission sent a reasoned opinion to Denmark for failure to inspect, given catches in excess of quotas, in 1988, 1991, 1992, 1994, 1995, and 1996. (A reasoned opinion is a written explanation of a decision, along with the reasons for that decision.)[75] The United Kingdom had excesses over certain quotas every year from 1991 through 1996; thus, action was taken in the Court of Justice against the UK. France similarly had excesses over certain quotas from 1991 through 1996, and court proceedings were begun.[76] While many of the infringement cases are settled before or shortly after the court's decisions have been made, it is significant that the EU does suffer from so many cases of lack of transposition or inappropriate application of EU directives. Although we will provide a comparison of the European Union and the United States later, it is worth noting here that non-compliance to federal environmental directives is not uncommon among states in the United States.

MONITORING EFFECTIVENESS

In order to monitor the effectiveness of the application of EU environmental legislation, the IMPEL (European Union Network for the Implementation and Enforcement of Environmental Law) network was created in 1992. IMPEL is made up of regulators of the member states of the European Union and focuses on the implementation and enforcement of environmental law. Its mission statement is to "protect the environment by the effective implementation of European Environmental law."[77] Its objectives are to work toward more effective application of environmental legislation through the "exchange of information and experience and development of a greater consistency of approach in the implementation, application, and enforcement of environmental legislation, with a special emphasis on Community environmental legislation."[78] The IMPEL network is not an official agency of the EU, rather it is an informal network generated at an informal meeting of the Community Environment Ministers in 1991. At that time, it was recognized that it would be advantageous to establish a network of representatives of relevant national authorities and the commission who would focus on the exchange of information and the development of practical common approaches to compliance and enforcement.[79] IMPEL is made up of the environmental authorities of EU member states and the European Commission, which shares the chairmanship at management meetings.[80]

A similar network of environmental coordinators from the accession countries was created to support those countries in meeting the implementation and enforcement obligations required of them to join the union. As of January 2003, IMPEL merged with the accession countries' IMPEL network (AC-IMPEL) to form one network, which now consists of the member states of the EU plus those countries that are applying for membership, a total of twenty-nine countries.[81]

IMPEL has published numerous reference materials for environmental inspections and compliance monitoring. The European parliament and council recommendation for providing minimum criteria for environmental inspections led IMPEL to develop a voluntary scheme for reporting and offering advice on inspection procedures.[82] By 2003, IMPEL had carried out five voluntary reviews of practices and procedures in environmental inspectorates.[83] The Irish review, completed in 2002, noted, for example, that arrangements for environmental inspections were in line with recommendations. It was suggested that Ireland's Environmental Protection Agency consider producing policy statements covering permitting, enforcement, and the handling of public complaints.[84]

The biggest obstacle to environmental policy success in the EU appears to be ensuring that environmental policy directives are carried out appropriately in the member states. (As we shall see subsequently, there are similar implementation problems in the United States.) The European Environment Agency's *Third Assessment* indicates that the major task facing the EU in the early 2000s is the implementation and enforcement of environmental policies. The report notes that there have been substantial improvements in some areas, such as increases in air quality and biodiversity, but most of the improvement has been through the traditional avenues of regulation. The report concluded that "the overall picture of Europe's environment remains complex."[85]

SUMMARY

In this chapter we have discussed several typical environmental situations faced in the EU. As we have seen, policies have sometimes been quite successful; in other cases, policies have not been as successful. We have illustrated situations in which positive and negative economic incentives have been employed to encourage the desired responses on the part of EU citizens. We also found the member states have been addressed certain environmental issues as members of the regime that had not been addressed by them as individual member states. In the next chapter we develop our conclusions, focusing on how our analysis of the EU can be applied to the development of internal environmental regimes.

CHAPTER 5

A COMPARATIVE EVALUATION OF THE EUROPEAN UNION AND THE UNITED STATES

INTRODUCTION

In this chapter we examine the European Union's environmental policies from a comparative perspective. We evaluate the similarities and differences between the European Union and the United States. The United States has a longer history as a union, but has not addressed environmental sustainability to the same degree as the European Union.

THE UNITED STATES ENVIRONMENT

The environment of the United States is similar to that of the European Union in many ways. As in the European Union, the land area of the United States includes deserts and forests, mountains and prairies, rivers and lakes. The landmass of the United States is roughly comparable to that of the EU, with the United States consisting of approximately 9,629,000 square kilometers (3,717,796 square miles) and the EU approximately 9,842,000 square kilometers (3,800,000 square miles).[1] The United States had, in the year 2000, 284 million citizens; the EU had 376 million.[2] Both have heavily industrialized regions and areas with such installations as nuclear power plants, as well as large agricultural belts. Both have a variety of laws covering various aspects of the environment and have regulations in place yet still face environmental challenges.

This chapter compares the European Union and the United States in their approaches to environmental policy. As noted in chapter 1, the norms and rules of a regime are important in influencing behaviors; thus, in this chapter we consider the approaches of the EU and the United States to discover how the approaches have influenced behavior related to the environment. First, we begin with an overview of the policy-making structure in the United States, with emphasis on those aspects that affect environmental policy. Then we turn to two specific cases in which the United States and the EU are different. In our first comparison, we examine how the United States and the European Union have approached the problem of global warming. Then we turn to a different environmental issue, that of environmental damage caused by contamination, examining the structures in place in the United States compared to the EU.

Approaches to Policy Making in the United States

The legal foundation for environmental law in the United States is found in Article I, section 8 of the U.S. Constitution. Known as the Commerce Clause, this provision gives the federal government the power to regulate interstate commerce. Over the past two hundred years, the powers of this clause have been interpreted so as to broaden its scope to include environmental matters.

Environmental law in the United States is primarily concerned with the allocation of resources, basically prevention of the dispersal of pollution into the natural environment and preventing the damage and depletion of natural resources. The U.S. justice system is based on the principles of English Common Law, with the concept of private property being preeminent. Traditionally the preeminence of private property provides the property owner with the right to pollute his or her land as much as he or she wishes, provided that the pollution does not extend into another's private property. Additionally, private property traditions allow the owner to use his or her property as desired. Thus, the landowner may pump as much water as desired from his or her land, regardless of the effect this may have on neighboring property owners. Most environmental protection measures are designed to place limits on the unconditional rights of the property owner.

In the United States, the foundation of environmental law and regulation rests on three basic mechanisms. First is the ability of Congress to pass laws. Second, the executive has the power to execute these laws through the formulation and implementation of regulations. The Administrative Procedures Act (APA) is the primary ruling governing the formulation and administration of regulations. The APA provides the framework for creating and implementing regulations, as well as other rule-making capabilities. The third mechanism on which environmental law and regulation are based consists of the methods provided for resolving disputes arising from the execution of policies and directives.

The adversarial justice system used in the United States is also based on English Common Law. This system requires that the two parties in a dispute bring their own supporting information to the court. However, in many cases related to the environment, the issue may involve those whose rights are not clearly defined. This is an important consideration because judicial standing, which determines who can and cannot bring a suit before the court, requires that the plaintiff prove a concrete interest in the matter before the court. Previous court cases involving environmental protection have established that judicial standing requires that the plaintiff must prove imminent injury in fact, that is, that the plaintiff was personally injured by the action in question. In addition, judicial standing requires that the court be able to remedy the situation in some way.

Environmental interest groups in the United States frequently fight their battles in the courts as they often lack the political resources needed to exercise influence in the legislative or executive branches of government. Since the early 1970s, a common legal challenge used by environmental organizations has been to enforce the environmental impact statement requirements of the National Environmental Policy Act (NEPA). In most cases, environmental impact statements are challenged on the grounds that they do not sufficiently assess environmental impacts or their alternatives. Both NEPA and the Clean Water Act have provided environmentalists with the ammunition they need to use the judicial branch of government to enforce environmental laws. As noted above, legal standing requirements often hampered environmentalists from seeking redress from the courts. However, since the early 1970s, the Supreme Court has been taking a more liberal view of standing in cases involving environmental issues.

Within the legislative branch, committees and subcommittees do the majority of work. This committee system makes it much more difficult to pass legislation than to block it. Consequently, those interests that are benefited by the status quo have an advantage in the policy-making process in the United States. Because there are so many different points in the legislative process where a bill can be defeated, those attempting to influence legislation must follow it closely through each stage in the process. Some groups have the resources to closely monitor legislation; others do not. In addition, those interest groups with a large and motivated membership are more likely to be influential in the legislative policy-making process than groups without these resources as the larger groups can enlist members to vote, campaign, and attend hearings.

The executive branch of government on the federal level is responsible for administering federal environmental regulations through a variety of federal agencies. The Environmental Protection Agency (EPA) is perhaps the most important regulatory agency concerned with environmental matters. The EPA was established under NEPA through a reorganization that consolidated pollution regulatory activities from a variety of different departments into a single agency. The EPA is responsible for a wide variety of environmental regulations and laws that provide the legal basis for its programs. The most well known of these are the Clean Air Act (1970); the Clean Water Act (1977); the Toxic Substances Control Act (1976); the Safe Drinking Water Act (1974); the Resource Conservation and Recovery Act (1976); and the Federal Insecticide, Fungicide, and Rodenticide Act (1972). Unlike the EU's European Environment Agency (discussed in chapter 3), which simply acts as a data-gathering and information-distributing agency, the EPA has regulatory authority as an agency. The EPA is authorized by the U.S. federal government to create regulations, that is, specific rules determining what is legal and what is not, and to enforce those regulations. For example, the EPA can regulate the level of a particular substance, for example, mercury, which is allowed under the Clean Water Act, and establish penalties for those entities that exceed the legal limit.

The administrator of the EPA is appointed by the president of the United States. Many EPA employees are highly trained technical experts, including engineers, environmental specialists, and other scientists. Others are professionals in other areas, such as legal and financial issues. The total EPA staff of eighteen thousand is spread over ten regional offices and seventeen labs located throughout the United States.[3]

The EPA produces five-year strategic plans, which outline ten long-term goals and the annual plans established to meet them. The strategic plan provides a basis for allocating resources to the highest priority issues. The 2000–2005 Strategic Plan includes goals related to air, water, food, and pollution.[4] Table 5.1 provides a list of the ten goals included in the EPA's 2000–2005 Strategic Plan.

The Interior Department is also involved in environmental matters, as is the Department of Agriculture. The agencies housed in these two departments are primary government players in the environmental policy implementation process. Agencies under the Interior Department include the Bureau of Land Management, the National Park Service, the U.S. Fish and Wildlife Service, and the Office of Surface Mining. The Department of Agriculture houses both the U.S. Forest Service and the Soil and Conservation Service.

In summary then, in contrast to the EU, the United States has a relatively centralized environmental administrative structure. When we compare the EPA in the Unites States with the European Environmental Agency, the U.S. EPA can make and enforce regulations and in that effort is backed by the executive and judicial branches of government. Yet the similarities of environment protection in Europe and the United States are greater than their differences. In both systems, environmental implementation is highly decentralized. The U.S. EPA depends on the state governments to develop state implementation plans (SIPs) to actually carry out on the state level the regulation that was developed on the national level. Although the term is not often used in the United States, the issue of subsidiarity is an important part of environmental regulatory policy in both systems. And, as we might expect, in both systems the units responsible for implementing environmental regulations (state governments in the United States and the member states in the EU) vary significantly in the rate and success of their environmental programs.

Table 5.1.
The U.S. Environmental Protection Agency: 2000–2005 Strategic Plan

The 2000–2005 Strategic Plan of the U.S. Environmental Protection Agency includes the following ten goals.[5]

1. Clean Air
   - Reducing air pollution in every community
2. Clean and Safe Water
   - Ensuring safe drinking water and clean waterways
3. Safe Food
   - Freedom from unsafe pesticides
4. Preventing Pollution and Reducing Risk in Communities, Homes, Workplaces, and Ecosystems
   - Safeguard natural ecosystems
5. Better Waste Management, Restoration of Contaminated Waste Sites, and Emergency Response
   - Appropriate treatment, storage and disposal of wastes to protect the environment
6. Reduction of Global and Cross-Border Environmental Risks
   - Lead other nations to reduce the risks associated with global environmental issues, including climate change
7. Quality Environmental Information
   - Provide public access to information concerning the environment and human health
8. Sound Science, Improved Understanding of Environmental Risk, and Greater Innovation to Address Environmental Problems
   - Apply the best available science in addressing environmental concerns
9. A Credible Deterrent to Pollution and Greater Compliance with the Law
   - Ensure full compliance with the law
10. Effective Management

Examination of Progress in the United States for the Past Twenty Years

In the past twenty years, there have been several subtle shifts in the emphasis and approach used in U.S. environmental policy. Like the EU to some extent, the United States has moved toward more market-based instruments for environmental protection. In 1990, amendments to the Clean Air Act created the first sulfur dioxide allowance trading program. Information programs, designed to increase transparency in environmental regulations, including the Toxics Release Inventory, have expanded. Issues of environmental justice, that is, concern over the disproportional impact of environmental harms on the poor and racial minorities, have become more prominent, and the use of benefit-cost analysis has expanded moderately. The last twenty years have also witnessed a large increase in concern about climate change.

Under President Ronald Reagan, the 1980s were years in which the administration wanted to dismantle many of the environmental regulations in the United States. Some segments of the public, however, reacted against the perceived antienvironmentalism of the Reagan administration, and membership in environmental organizations soared. A number of new types of environmental groups also arose during this period, including more radical groups such as Earth First.

After the Reagan presidency, the approach toward environmental matters taken by the next two presidents, Bush and Clinton, was much less confrontational. President Bush went in a new policy direction from his predecessor, overseeing many environmentally important efforts, such as amending and reauthorizing the Clean Air Act. The Clinton administration placed strong environmental advocates in important positions, appointing Bruce Babbitt as secretary of the interior and Carolyn Browner as director of the Environmental Protection Agency. Clinton's vice president, Al Gore, began a worldwide program for school children aimed at increasing their environmental awareness. The program, the Global Learning and Observations to Benefit the Environment Program (GLOBE), has students, under teacher supervision, measure and monitor environmental quality.[6] The views of the administration of President George W. Bush are described later.

On the part of the public, national polls in recent years indicate that concern for the environment seems to have leveled off during the past twenty years. It may be that the public perceives that the environmental policies have been successful and the danger to the environment and human health from polluting activities is not as significant as it once was.

With this brief discussion of U.S. policymaking covered, our next step in understanding environment regimes is to examine two environmental issues and the differing approaches the EU and the United States have taken on these. We will also notice that the EU has at times based its environmental policy formation on what has happened in the United States.

GLOBAL WARMING

One of today's major environmental concerns is global warming. (While many readers are familiar with this problem, we will provide some background information for those who may not be as knowledgeable.) Global warming is the increase in the average temperature of the earth's atmosphere. Many gases that naturally occur in the earth's atmosphere, such as carbon dioxide and water vapor, tend to trap heat. (The atmosphere consists mostly of oxygen and nitrogen, 21 percent and 78 percent, respectively.)[7] The trapping of heat in the atmosphere occurs naturally to some degree. As heat energy from the earth rises, it is absorbed by water vapor, carbon dioxide, and other natural greenhouse gases in the atmosphere. The term *greenhouse gases* is used because of the effect of the atmosphere in keeping the earth warmer than it would otherwise be if the earth's heat energy went straight out into space (in that case, the earth would be too cold to be habitable). However, the levels of heat-trapping gases in the atmosphere have increased substantially with industrial production and the burning of fossil fuels so that the balance of incoming and outgoing energy in the earth's atmosphere has become unbalanced. As a result, the earth's surface temperature is warming, affecting the global climate.

The gases associated with increased global warming include methane (CH_4), chlorofluorocarbons (CFCs) and their substitutes, and nitrous oxides (associated with fertilizer use). These gases are referred to as "stock pollutants" because they remain in the atmosphere for long periods, accumulating over time and mixing with other gases in the atmosphere. As a result of the atmospheric mixing of gases, the atmospheric impact of the stock pollutants is distributed throughout the world. These stock pollutants, while in lower quantities than CO_2 in the atmosphere, hold much more heat and are significant greenhouse gases. Thus, the level of heat trapped in the atmosphere has been increasing, hence the term *greenhouse effect.*

The greenhouse effect is associated with a number of potential climatic problems including an increase in severe weather, decreased biodiversity, changes in drought and flood patterns, and rising sea levels. Historically, the earth's climate has changed, but because of human activity, the rate of change today is significant. Over the last one hundred years, the earth's temperature has risen 1 degree Fahrenheit; this is associated with an increase in the concentration of greenhouse gases by about 30 percent.[8] While the cause of global warming is still being debated, there is reason for prudence. Both the Intergovernmental Panel on Climate Change (IPCC) and the National

Research Council have concluded that evidence indicates the influence of human activity on the global climate.[9] (The IPCC was set up by the World Meteorological Organization and the United Nations Environment Program in 1988, the same year in which the United Nations General Assembly adopted resolution 43/53, on the Protection of Global Climate for Present and Future Generations of Mankind.) The IPCC projects that, by 2100, the global surface temperatures will rise by 1.4 to 5.8 degrees Centigrade, the most rapid increase since the last ice age.[10] In 1992, at the Rio Earth Summit, the United Nations Framework Convention on Climate Change was developed to address the important issue of climate change.

The Framework Convention on Climate Change focuses on climate change and reducing the quantity of greenhouse gases released into the atmosphere. The convention, which took effect in 1994, has been ratified by 186 countries. (Ratification means that the states are legally bound to the stipulations of the treaty. Those countries that ratify are referred to as "parties to the convention.") The convention set objectives for the stabilization of greenhouse gas concentrations at a level that would prevent human actions from dangerously altering the earth's climate. The convention calls for parties to "commit to the overall goal of stabilization of greenhouse gas concentrations in the atmosphere at a level that would prevent dangerous anthropogenic interference with the climate system... within a time frame sufficient to enable economic development to proceed in a sustainable manner."[11] The time frame is imprecise because scientific knowledge at this point is uncertain. Although scientists generally agree that climate change is a potentially serious problem, the magnitude of its effects and when and where they might occur has not yet been ascertained. Thus the treaty promotes the precautionary principle; that is, in cases of scientific uncertainty, the safest strategy is to restrict or prohibit activities that could potentially cause serious harm.

The parties to the Climate Convention agree to take action in their countries to slow climate change. The convention also encourages the sharing of technology and cooperation on reducing greenhouse gas emissions. Each party to the conventions is required to develop an inventory of sources of greenhouse gases, such as, factories, as well as sinks (areas such as forests that absorb greenhouse gases) within the country. This data is used to measure emission changes and to develop appropriate strategies for reducing emissions. (It should be noted that there are no enforcement mechanisms associated with the convention.)

The Framework Convention on Climate Change was designed to allow countries to adjust the treaty as appropriate when new science develops through the adoption of protocols to the convention. (A protocol is like an amendment and is linked to an existing treaty but is considered an international agreement that can stand on its own.) The Kyoto Protocol is the most well known of these. Developed at the Conference of Parties held in Kyoto, Japan, in 1997, the Kyoto Protocol calls for a number of emission reductions designed to decrease greenhouse gases in the atmosphere and requires developed nations to decrease their emissions of greenhouse gases.

While the Kyoto Protocol developed basic features of the system, it did not provide operational rules. Additional Conferences of the Parties were held to further develop the features of the protocol. The design of the system was solidified over a number of meetings held in Buenos Aires (resulting in the Buenos Aires Plan of Action) in 1998, Bonn, Germany, in July 2001 (the Bonn Agreements), and in Marrakesh in October/November 2001 (the Marrakesh Accords). The Marrakesh Accords contain the rulebook for the Kyoto Protocol as well as improvements for implementing the convention.

With this background and understanding of global warming and the Kyoto Protocol in place, we now examine the positions of the United States and the EU on climate change.

Global Warming: Comparative Perspectives

On a per capita basis, the United States is the largest producer of greenhouse gases in the world, producing 6.6 tons of greenhouse gases per person every year.[12] The majority of the gases (approximately 82 percent) are caused by the burning of fossil fuels for electricity and transportation. U.S. emissions actually increased by about 3.4 percent per person between 1990 and 1997; however, by 2000, various strategies to reduce emissions led to an estimated decrease of 2.7 percent of total emissions.[13]

The United States has not been supportive of the Kyoto Protocol. President George W. Bush has voiced his opinion that the Kyoto Protocol is fundamentally flawed because it "does not establish a long-term goal based on science."[14] President Bush has also argued that the protocol poses "serious and unnecessary risks to the US and world economies" and that signing on to the Kyoto Protocol would "put millions of Americans out of work and undermine our ability to make long-term investments in clean energy."[15]

The emission reduction target for the United States under the Kyoto Protocol would be 7 percent per year from 2008 to 2012. The Bush administration considers this figure to be deceiving as it does not consider increases in emissions from 1990 through 2012; the actual decrease is estimated to be approximately 30 percent. The administration argues that such a large decrease in emissions would require great sacrifices in the economy of the United States. In addition, because the Kyoto Protocol includes an emissions trading system, the United States would need to trade with other countries in order to meet its reduction targets. The Bush administration does not support this structure as there is no guarantee the United States would be able to make the needed trades, which might, in any event, be prohibitively costly. The administration also objects to exemptions provided to developing nations. In addition, the administration argues that the Kyoto Protocol does not appropriately address substances such as black carbon and tropospheric ozone, which impact the climate and pose significant health risks.[16] In light of its disagreement with these requirements, the United States did not sign the protocol.

In place of the demands of the Kyoto Protocol, the Bush administration has suggested alternative approaches to address climate change. The Climate Change Technology Initiative, for example, includes the development of technologies for measuring and monitoring greenhouse gas emissions and the development of private-public partnerships in research and development efforts designed to discover cost-effective ways to reduce greenhouse emissions.[17]

The U.S. alternatives to the Kyoto Protocol's approach to combat global warming include the Clear Skies and Global Climate Change Initiatives, announced by President George W. Bush in 2002.[18] The Clear Skies Initiative is designed to cut power plant emissions of nitrogen oxides by 67 percent, sulfur dioxide by 73 percent, and mercury by 69 percent. The Global Climate Change Initiative presents a strategy to decrease greenhouse gases by 18 percent over the next ten years. The initiative also supports research on climate change by providing $4.5 billion in the president's fiscal year 2003 budget for climate change–related activities and includes the first year's funding for a five-year $4.6 billion commitment to tax credits for renewable energy sources.

In contrast to the United States leaders, EU leaders have been very concerned with solving the problem of global warming and supportive of the Kyoto Protocol. In 2002, European Environment Commissioner Margot Wallström described climate change as the main environmental problem today.[19] Commissioner Wallström notes that the EU is committed to developing strategies to avert climate change. Limiting climate change is one of the EU's strategies for sustainable development, which will be implemented through the commitments of the Kyoto Protocol. The EU goal is to cut "greenhouse gas emissions by an average of one percent per year over 1990 levels through the year 2020."[20] The EU Heads of State and Government reiterated this commitment at the European Council in June 2001, stating that combating climate change is one of the priorities of the EU's Sustainable Development Strategy.[21]

The EU has taken the position espoused by the precautionary principle: it is safer to take action while scientific understanding is improved than to wait until certainty is achieved because waiting may result in irreparable damage; thus, the EU took a strong and leading position on the United Nations Framework Convention on Climate Change. The convention requires industrialized nations to assist developing nations meet their obligations by providing financial help and technology transfers, a requirement to which the EU has committed itself.

The EU has integrated climate change concerns into many areas, in line with its commitment to sustainable development. The Sixth Environmental Action Program includes the establishment of an EU-wide CO_2 emissions trading scheme as one of its objectives.[22] (As in the United States, carbon dioxide (CO_2) represents about 82 percent of total global warming gases emitted in the EU.)[23] In addition, the "Green Paper on the Security of Energy Supply" places work in the area of global warming on a high priority, as does the "White Paper on a Common Transport Policy." Both papers encourage strategies designed to more efficient use of energy. The European Climate Change Program (ECCP) was established in 2001 with the specific purpose of identifying the most environmentally and cost effective approaches to meeting the EU targets for reductions in greenhouse gases.

Like the EU, the United States is involved in climate change research. The U.S. Global Change Research Program (USGCRP) coordinates much of the federal research on climate change, in which ten different federal agencies are involved. Since its establishment under the Global Change Research Act of 1990, the USGCRP has spent approximately $1.6 billion annually in climate change research.[24] About half the annual expenditures are on satellite systems; the other half are dedicated to climate change science. USGCRP research has included research on human linkages to ozone depletion, climate change, and land cover changes.

As we have seen, both the United States and the EU declare that they are concerned about global warming and both have established research programs to study the problem. The EU, on the other hand, has gone much further, accepting that global cooperation is needed and supporting the Kyoto Protocol as the means of achieving the goals of reduction of greenhouse emissions. The United States, on the other hand, has chosen to work on its own, in fear of potential economic loses associated with reductions in greenhouse emissions. The United States does not have an overall sustainability objective, as does the EU. This guiding principle has had a significant effect on the approaches the EU has taken, leading the EU to integrate greenhouse gas reduction policies into objectives in many areas. Because of the orientation of the EU, we would expect much more progress to be made in greenhouse gas reduction by 2012, compared to progress made by the United States, where economic concerns are given priority.

The case of global warming provides an example of the differences between the United States and the EU in dealing with a politically external environmental problem. The responses of each system reflect the politics of their respective systems. For example, in the United States the change in a national administration combined with the ability of the president to represent the United States internationally without input from other parties allows one view to predominate—even when that position is different from the U.S. position a few years earlier. In contrast the EU system requires consensus building over time by various parties and interests. We feel that cultural differences also play a role in the differing approaches of the United States and the EU. Specifically, the more communitarian social orientation of some European societies—notable in northern Europe—contrasts with the more individualistic lassiez-fare orientation that is associated with the United States.

ENVIRONMENTAL LIABILITY

We now turn to another environmental concern: who is responsible for remedying the situation in the case of large-scale toxic contamination of land or water resources? Numerous environmentally devastating events have occurred over the years, affecting large areas of land, water resources, and/or wildlife. For example, damaged and wrecked oil tankers have become notorious for the damage they cause to water life, bird life, and coastal lands. Such events severely tax the financial and physical resources of a nation to clean up and repair the damage. In this section, we examine how the United States and the EU have dealt with such occurrences. We discuss who pays for the damages and who is responsible for enforcing environmental laws and preventing such incidents. As we will see, while the United States has had regulatory structures in place for three decades; the EU is only recently acknowledging a need to address such issues.

Environmental Liability: Comparative Approaches

In the 1960s and 1970s, contaminated sites such as Love Canal, where people were living with toxic substances, and Rachel Carson's book, *Silent Spring*, decrying the dangers of toxic chemicals in the environment, brought public attention to the environment. In the following years, numerous environmental laws were put in effect in the United States, as we have noted above.

In the United States, liability for environmental damage is addressed in the Comprehensive Environmental Response, Compensation, and Liability Act (CERCLA, also referred to as Superfund), enacted in 1980. CERCLA provides for federal authorities to respond to releases or threatened releases of hazardous substances. The U.S. Environmental Protection Agency (EPA) is the agency charged with identifying contaminated sites. Under CERCLA, the EPA may undertake cleanup of contaminated sites, or it may force the responsible parties to either pay for the cleanup or undertake the cleanup directly. The act also provided for a tax on chemical and petroleum industries, which netted a fund (placed in trust) of $1.6 billion over five years and was designated to cover the costs of restoring sites when the responsible party cannot be identified or when the responsible party cannot pay the cleanup costs (hence the term *Superfund*). Superfund sites are not limited to the fifty U.S. states but also include areas in Puerto Rico, Guam, American Samoa, the U.S. Virgin Islands, the Commonwealth of the Northern Mariana Islands, and U.S. trust territories. Many of the sites both outside and within the fifty U.S. states are former military installations. Figure 5.1 provides brief descriptions of two Superfund sites.

According to CERCLA, two forms of response, to either the threat of release or the certain release of hazardous substances, are authorized. First, actions may be taken quickly when needed to respond to releases or threatened releases. Second, remedial actions may be taken to respond to serious, but not immediately life-threatening, releases or threats of releases of hazardous substances. Remedial actions may only be taken at sites listed on the EPA's National Priorities List (NPL).[25] To be placed on the NPL, the site must first be proposed in the *Federal Register*, after which the EPA accepts and responds to public comments on the proposed sites. In order to be eligible to be placed on the NPC, three possible means are available. First, the EPA's Hazard Ranking System (HRS) provides an avenue by which sites may be placed on the NPL. The HRS is a numerically based system that uses preliminary assessments to evaluate the potential harm to human health or to the environment posed by sites.[26] The factors considered in ranking sites under the HRS include the characteristics of the waste (e.g., the level of toxicity), whether people or sensitive environments are affected, and the likelihood that hazardous substances have or will be at the site. Second, U.S. states or territories are allowed to designate one top-priority site regardless of score. Third, if a site meets three specific requirements, it may be placed on the NPL. These three requirements include (1) The U.S. Public Health Service's Agency for Toxic Substances and Disease Registry issuing a health advisory recommending the removal of people from the site, (2) the EPA determines that the site is a significant threat to public health, and (3) the EPA expects that remedial action will be more cost effective than would using emergency removal authority at the site.[27] Since its inception in 1980, over 1,370 Superfund sites have been listed, with potentially more than 3,000 additional sites to be evaluated.[28]

The Superfund Amendments and Reauthorization Act (SARA), 1986, amended CERCLA, making additions and changes to the original act. Changes included increasing the size of the trust fund to $8.5 billion, encouraging greater citizen participation in site cleanup, providing new enforcement authorities, and increasing state involvement. SARA also required the EPA to make certain that the Hazard Ranking System accurately evaluates the degree of risk to human health and the environment.[29]

In addition to CERCLA and SARA, the United States has additional regulations covering certain hazardous substances. These include the National Oil and Hazardous Substances Pollution Contingency Plan (NCP) to respond to oil spills and the release of hazardous substances on a national level. First developed in 1968, the plan has been broadened over the years to cover hazardous substance spills and releases at hazardous waste sites, which require emergency removal actions under Superfund. More recently, the oil spill provisions contained in the Oil Pollution Act of 1990 resulted in revisions to the NCP in 1994 to reflect those requirements.[30]

In contrast to the United States, the EU has only more recently addressed the problem of environmental liability. Environmental disasters such as the *Erika* tanker oil spill off the coast of France and chemical contamination of the Doñana National Park in southern Spain called attention to the need for legislation regarding environmental damage in the EU. (See Figure 5.2 for a brief description of these two incidents.) While the commission and governments of member states have cooperated in response to such incidents (the magnitude of many such environmental events is often more than one country can handle alone), there is no EU directive covering actions in regard to large-scale environmentally disastrous events. (As in the United States, the EU has civil protection in place to protect workers and ensure the safety of manufactured goods.)

As noted previously, the EU follows the polluter pays principle, that is, the party responsible for the polluting activity is responsible for rectifying the damage due to that pollution. While member states have national liability regulations in place to cover damage to people and property, most of these do not extend to environmental damages. To remedy this situation, in 2002 the European Commission proposed a directive on environmental liability, the *Proposal for a Directive of the European Parliament and of the Council on Environmental Liability with Regard to the Prevention and Remedying of Environmental Damage*.[31] This proposal follows a Commission White Paper in 2000 on environmental liability and a Commission Sustainable Development Strategy, which suggested that legislation on environmental liability should be in place in the EU by 2003.

The environmental liability proposal is designed to establish a framework for the prevention or remedy of environmental damage. Environmental damage is defined "in reference to biodiversity protect at Community and national levels, waters covered by the Water Framework Directive and human health when the source of the threat to human health is land contami-

Figure 5.1.
Two U.S. Superfund Sites

The Forest Glen Mobile Home Subdivision

The Superfund's Forest Glen site is located between the City of Niagara Falls and the Town of Niagara, New York. In the 1970s, the site was used as an illegal waste dump. The toxic substances dumped there, including polycyclic aromatic hydrocarbons and semi-volatile organic compounds, were covered with a thin layer of topsoil, and the land was sold. In 1980, the Niagara County Health Department discovered the contaminated soil; the EPA became involved in 1987. The contaminated area was fenced off and by 1992, 53 families had been permanently relocated.

Four parties responsible for the dumping are currently involved in the cleanup. A settlement between Goodyear Tire and Rubber Company and the EPA has Goodyear paying the costs of cleanup, estimated at $16 million. Goodyear will also reimburse the agency $9 million for costs already incurred by the EPA and damages to natural resources. The other three responsible parties agreed to pay $81,000 to reimburse the agency for costs incurred. Cleanup efforts for the site have included the treatment of 125 million gallons of groundwater and 18,000 cubic yards of soil.

The Eastland Woolen Mill

The Eastland Woolen Mill was placed on the EPA's National Priorities List in 1999. Located in Corinna, Maine, the mill was a major employer in the town until it declared bankruptcy and closed in 1996. During its operation, the mill had dumped toxic substances into floor drains which emptied into the soil and the Sebasticook River. Drinking water that had been contaminated by chlorinated benzene compounds was discovered in 1983, forcing water supplies to be fitted with special carbon filters. The EPA discovered the toxic compounds in the soil, groundwater, surface water, and sediment, from which it had spread to the town's water supplies.

Because the mill had been located in the center of Corrina, the main street had to be relocated, a road replaced, and 115,000 tons of contaminated soil removed. The EPA incurred costs of $36 million between 1999 and 2002; soil treatment, however, will continue until 2004.

Sources: U.S. Environmental Protection Agency. "Success Stories—Eastland Woolen Mill." Accessed 10 November 2002. Available from http://www.epa.gov/superfund/-action/success.htm; U.S. Environmental Protection Agency, "Success Stories—Forest Glen Mobile Home Subdivision." Accessed 10 November 2002. Available from http://www.epa_gov/superfund/action/success.htm.

nation," and is based on Article 175(1) of the EC Treaty.[32] (The Water Framework Directive expands the scope of water protection to all waters and focuses on an approach combining emission limits and quality standards.)[33] The environmental liability proposal states that there is a need for such a directive because there are already an estimated three hundred thousand sites throughout the EU that have "been identified as definitely or potentially contaminated."[34] The proposal notes that most of the damage is the result of past occurrences, in a time before most member states had developed regulations and penalties for environmental liability. Thus, it may be impossible to seek payment from the offenders, necessitating the spending of public (EU) monies in cleanup efforts. In addition, not all member states have adopted legislation: among those that have, legislation often does not require the clean up of orphan sites (those sites where the party responsible for the damage cannot be identified, or in which the responsible parties are not financially viable, e.g., bankrupt). In addition, a commission study concluded that, for liability measures to be effective, they should be consistently applied at the community level.[35]

The environmental liability proposal is designed to create a common base among member states by setting rules on restoration objectives and appropriate actions. Thus the proposal requires member states to ensure that environmental damage is remedied. Based on the polluter pays principle, the state authority is to recover restoration costs from the responsible parties. The determination of when the responsible party, the appropriate authorities, or even a third party, is required to take measures to prevent or remedy environmental damage would be left to the discretion of the member state, in line with the subsidiarity principle (discussed in chapter 2). Provisions do provide for action by "qualified entities, alongside those persons who have a sufficient interest, to request the competent authority to take appropriate action and possibly challenge their subsequent action or inaction."[36] This suggests that NGOs or members of the public would be able to demand action when environmental damage occurs.

Like industry stakeholders in the United States, those in the EU have argued that the proposed directive goes too far and have requested that the proposed directive include a ceiling on liability payments. However, EU Environment Commissioner Margot Wallström has warned that ceilings would alleviate the need for industry to be concerned about the amount of damage caused.[37] Wallström contends that stricter liability regulations will force industry to be more environmentally responsible. In contrast to the industry view, environmental nongovernmental organizations (ENGOs) in the EU have argued that the directive is not strong enough.

Figure 5.2.
Two Incidents of Environmental Damage in the EU

The Doñana National Park, Spain

The Doñana National Park, located in the Andalusia area of southern Spain, is Europe's largest nature preserve. Founded in 1969, the park covers 180,000 acres. The park contains three ecosystems: wetlands, pine forest, and sand dunes and is the home of hundreds of permanent and migrating birds, including five threatened species.[38] Some of the last surviving lynxes in Europe as well as red deer also make their homes in the park.[39]

In 1998 a dike downstream from a Canadian-owned minerals plant collapsed, sending a stream of water contaminated by zinc, lead, and cadmium into the Guadiamar River, which borders the park. Greenpeace and other environmental groups expressed outraged over the incident.[40] Great Britain aided France in bird rescue and cleanup efforts.

The *Erika* Tanker Oil Spill, France

In 1999, the oil tanker *Erika* broke in two off the coast of France, spilling 11 million tons of oil into the sea. Large stretches of coastland were covered by oil, and at least twenty-one thousand seabirds killed by the disaster and miles of beaches closed.[41] Besides the environmental toll on birdlife, water creatures were also affected. France recommended avoiding shellfish, oysters, and mussels from the polluted area.

TotalFina, the French firm that had chartered the tanker, offered to pay over $60 million for cleanup efforts.

As we have seen, in both the United States and the EU, the impetus for liability regulation is protection of the environment and of human health. Under both Superfund and the EU proposal, responsible parties are to be held liable for environmental damages. The appropriate authorities are responsible for remedying the damage. However in the United States there is one national entity, the EPA, that is given primary responsibility for identifying sites and taking steps to remedy the situation. In the EU, there are entities at the national level in each member state that have varying degrees of authority. In addition, the subsidiarity principle provides for each state some latitude as to how to approach the problem. This situation could potentially result in unequal application of the directive and inconsistent remedial action among the EU member states.

In the United States, CERCLA has a well-established protocol for ranking and handling contaminated sites. In the EU, similar levels of data have not been gathered about the sites that have been identified, making estimates of cleanup costs less predictable at this point. CERCLA also requires responsible parties to be liable for contamination, even if the contamination occurred before the law came into force. The EU proposal does not stipulate a similar policy for previous damage. In fact, as noted in the proposal, the commission expects that the EU will have to provide funding for at least part of the cleanup efforts of these sites.

In the United States, CERCLA regulations focus on the particular hazardous substances associated with each site. However, the EU proposal focuses on the activities (rather than substances) that cause environmental damage. While, of course, damage-causing activities are related to the damage-causing substances, the EU approach is a bit broader in scope. By focusing on activities, rather than just substances, the EU approach can factor in the effects of human activity on, for example, biodiversity, which the U.S. approach would not ordinarily capture.

A final difference is the broad range of parties who can be held responsible under CERCLA, compared to the EU proposal. Under CERCLA, any combination of companies and/or individuals, generators or transporters of hazardous substances, and operators of waste disposal sites can be held responsible. This allows the liability to be spread among several parties and increases the possible recovery of costs expended by the EPA in cleanup efforts. In the EU, only the operator would be held responsible. This could severely limit the possibilities of recovering large costs from the responsible operator and require a larger proportion of cleanup costs to be covered by member state or EU funding.

We offer that some of the differences between the U.S. and the EU liability systems can be accounted for by the context in which each system was developed. In the U.S. CERCLA was passed during an ambitious period of relatively high environmental awareness on the national level. In the United States policy is often not made or significantly changed, unless there is a "crisis" atmosphere. The results can be both good and bad. Sometimes the crisis atmosphere produces outcomes that would otherwise not be possible. In contrast, normally, in the EU the multiple layers of policy-making make "crisis" policy formation possible but less likely. In our opinion focusing liability on the industrial process involved and the producer is a better approach, and we suspect the fact that it was not developed rapidly or under crisis conditions contributed to the reasoned development of this policy in the EU.

HAZARDOUS WASTE

Toxic wastes are a by-product of energy development, agriculture, and most industrial activity. They are found throughout the environment, in the air, water, and soil.

Hazardous Waste: Comparative Approaches

In the United States, toxic substances are regulated under a variety of federal acts. Chemicals used commercially are regulated under the Toxic Substances Control Act, the Federal Insecticide, Fungicide, and Rodenticide Act, and the Food, Drug, and Cosmetics Act. These acts require that the manufacturers of chemicals conduct their own product safety tests and submit the results to the federal government. The regulatory agency involved then makes a decision on a case-by-case basis as to whether or not to ban the chemical, regulate its use, or leave it unregulated. The tests are also used to determine safe levels of exposure to chemical residues in drinking water, air, food, or the workplace.

Chemicals in the EU are governed under three directives and include both existing and new substances. Provision of data is covered under Directive 67/548, which encompasses new substances, and Regulation 793/93, for existing substances. Under this directive, manufacturers or importers must provide data on new substances before they can be marketed; substances already on the market can continue to be sold while the data is being provided. Directive 93/67 and Regulation 1488/94 cover risk-assessment procedures related to chemicals. Risk assessment is done under the auspices of the Directorate-General of the Environment and provides the basis for marketing or use restrictions. Directive 76/769 limits marketing and the use of dangerous substances.[42] EU legislation in this area has been criticized as not sufficiently covering hazardous substances; for example, in the EU there are currently 100,106 substances that can be used without testing, and the burden of proof of harm is on public authorities.[43]

The United States and the EU thus have similar approaches to the regulation of toxic chemicals. In both communities, there are problems with the legislation so that certain chemicals may not be sufficiently covered under current regulations.

SOLID WASTE

On a per capita basis, the United States generates more than twice as much municipal waste as most of the more developed countries in the world. About 70 percent of municipal solid waste is landfilled, about 17 percent is recycled, and the rest is destroyed through incineration (the fastest growing method of disposal).[44] The requirements of the Resource Conservation and Recovery Act (discussed below) have led to the closing of many landfills. It has been estimated that over 75 percent of surveyed states have insufficient solid waste landfill capacity.[45] Besides the space problems associated with landfills, improper disposal of solid waste can result in the pollution of nearby waters.

In 1965, the United States put into effect the Solid Waste Disposal Act. However, this legislation is only advisory; traditionally, solid waste disposal has been the responsibility of the state and local governments. It was not until the passage of the Resource Conservation and Recovery Act of 1976 (RCRA) that the federal government got involved to any significant degree in the management of solid waste. Although most of RCRA deals with the disposal of hazardous waste, it also created, for the first time, significant federal regulations for the management of municipal waste. The RCRA required states to develop solid waste management plans, which among other things would stipulate the closing of all open dumps and either the recycling of wastes or the disposal of waste within sanitary landfills.[46]

RCRA requires that anyone storing, treating, or disposing of hazardous waste do so under permit from the EPA. Permits are granted only to those firms that can demonstrate that they have the financial capacity, the necessary insurance, and the expertise to know how to operate a landfill or disposal facility. The RCRA applies to any producer of hazardous waste generating at least 220 pounds of waste a month, an amount that fills a 55-gallon drum about halfway. The act provides the EPA with the authority to impose fines and hold individuals criminally liable for the improper disposal of waste.

In the EU, much effort has been made to reduce solid waste. Council Directive 75/442/EEC, effective July 1975, and its amendments govern the system of waste management in the EU with the objective of limiting waste production. Under this directive, companies that treat, store, or dump waste must obtain permission from the competent authority. Requirements depend on the particular type of waste involved.[47]

The landfilling of waste in the EU is governed under a separate directive, Council Directive 99/31/EC on the landfill of waste. The directive classifies landfills according to whether they are for hazardous waste, nonhazardous waste, or inert waste; however, certain types of waste are prohibited from being placed in a landfill. Prohibited substances include items such as liquid waste, flammable waste, explosive waste, infectious waste, and used tires.[48]

Waste legislation in the EU is thus similar to that in the United States, but the EU has been more aggressive in many ways. For instance, there is one area, in particular, in which the EU is advancing beyond U.S. requirements, that of end-of-life vehicles. Directive 2000/53/EC on end-of-life vehicles designed to promote the collection, reuse, and recycling of the vehicle components. Anyone who has seen a used car junkyard is well aware of the waste that end-of-life vehicles represent. The directive requires that manufacturers of vehicles and vehicular equipment must avoid the use of hazardous substances when possible, design the vehicles to promote reuse and recycling of the component parts, and increase the use of recycled materials in vehicle manufacturing. Owners of the end-of-life vehicle are able to dispose of the vehicle for free (the "free take-back" principle); the vehicle's producer is required to cover the costs of this. In 2002, about 75 percent of end-of-life vehicles had their metal components recycled. The goal of the directive is to increase that average recovery (by weight per vehicle) to 85 percent by 2006, and to 95 percent by 2015.[49] This is obviously a very strong step made toward recycling and environmental protection, compared to the United States.

We suspect that part of the difference in waste disposal policies in this area can be attributed to the lack of landfill space in many parts of Europe. In the United States, as landfill space has become scarce, notably in the northeast, state and local governments have introduced recycling programs.

SOIL EROSION

Soil erosion is the movement of soil, either by wind or water, off farmlands and into lakes, rivers, streams, or the oceans. Soil erosion develops slowly and seemingly has little impact; however, over time its cumulative effects can be significant.

Soil Erosion: Comparative Approaches

In the United States, about 6 billion tons of topsoil are lost each year. Historically, this amounts to about one-third of the topsoil originally on U.S. croplands. In the United States, the Soil Conservation Service within the Department of Agriculture has primary authority over soil erosion. Although the service provides technical assistance and produces useful data, it has no authority to require agricultural practices that prevent erosion.

The EU has similar erosion problems, particularly in the Mediterranean states. About 12 percent of the total European land area is affected by water erosion and about 4 percent, by wind erosion.[50] Erosion has increased since 1950 because of increased mechanization, ploughing on steeper slopes, and the loss of grass cover due to overgrazing.

The EU's Common Agriculture Policy (CAP) has so far done little to address soil erosion and other environmental issues connected to agriculture; neither did the 1992 CAP reforms do much to integrate environmental concerns into agricultural policies. However, since the 1992 reforms, the commission has been taking steps toward integrating environmental concerns into agricultural policy in order to move toward a more sustainable agriculture. (The European Commission has referred to sustainable agriculture as "the management of natural resources in a way which ensures that their benefits are also available for the future.")[51] For example, one approach to reduce soil erosion is to limit the number of cattle units per hectare eligible for support payments; eligible cattle have decreased from 3.5 cattle in 1992 to 2 cattle in 1996.[52] In addition the Structural Funds rules governing investment funding now allow for a wider range of environmental investments in agriculture to help combat problems such as erosion.

One of the objectives of the Sixth Environmental Action Program is to protect soils against erosion and pollution. It is to fulfill this objective that the commission is publishing this communication, which paves the way for developing a strategy on soil protection.[53]

The EU, principally through the cattle program, is clearly more aggressively dealing with soil erosion than is the United States. However, given the level of soil erosion in both Europe and the United States one could reasonably conclude that neither system is dealing aggressively (or even adequately) with soil erosion. We suspect this is due, in part, to the influence and importance of the agricultural sector in the political systems of both the United States and the EU.

WATER POLLUTION

The United States has three major laws that regulate water pollution: the Clean Water Act, the Safe Drinking Water Act, and the Resource Conservation ad Recovery Act (discussed above). The Clean Water Act required that any discharge of a pollutant from a point source is allowed only pursuant to a permit issued by the EPA or by a state agency after EPA approval of a state plan. A point source is defined as "any discernable, confined and discreet conveyance, including but not limited to any pipe, ditch, channel, tunnel, conduit, well, discrete fissure, container, rolling stock, concentrated animal feeding operation, or vessel or floating craft from which pollutants are or may be discharged."[54]

Permits to pollute are issued under the Clean Water Act's National Pollution Discharge Elimination System. While over sixty-five thousand permits have been issued, the system has not lived up to its early promise. Budget cuts have led to the necessity that EPA and state officials rely on industry for information used in establishing allowable discharges under the permit system. In addition, poor staffing has resulted in haphazard enforcement, with action being taken against the most visible violators while other violators go undetected.

The Safe Drinking Water Act (SDWA) of 1974 and its 1986 amendments regulate drinking water produced by public water supply systems. The initial act required the monitoring and regulation of twenty-two different water contaminants. The amendments required that by 1991, eighty-three additional contaminants were to be added to those covered by the original act. An additional twenty-five contaminants are required to be added every three years. EPA regulations developed pursuant to the SDWA further specified criteria under which public water systems relying on surface water must install filtration equipment.

As with the Clean Air Act, the SDWA has suffered from enforcement problems. However, citizens may not file suit to force compliance. (Citizen suits allow any interested party to sue the polluter or even the EPA for failure to enforce the regulations.)

As in the United States, more than one piece of EU legislation applies to water. However, in the EU, as examined in chapter 4, water pollution legislation is not nearly as well developed as in the United States. The primary water directives in the EU are the Drinking Water Directive (80/778/EEC), the Commission Communication of February 1996 on Community Water Policy, and Council Directive 80/778/EEC relating to the quality of water intended for human consumption.

The European Commission's Communication on Community Water Policy "outlines the approach to water protection to be used in the community. It describes the objectives of community water policy. These include providing for a "secure supply of drinking water, secure supply of drinking water or non-drinking water to meet economic requirements other than human consumption, protection and preservation of the aquatic environment, and the restriction of natural disasters (drought, floods)."[55] The principles of the policy are essentially those underlying other recent EU legislation. These principles are listed in table 5.2.

The EU's Drinking Water Directive regulates the maximum concentrations of various substances allowed in water. Council Directive on the Quality of Water Intended for Human Consumption (OJ L330, 5.12.98) expands the scope of water protection to cover both surface waters and groundwaters and bases water management on river basins. Both emission limit values and quality standards are used to improve and maintain water quality.[56]

The EU has moved a step ahead of the United States in one particular aspect of water management, that of water pricing. While individual U.S. municipal water providers may at times, such as during droughts, be forced to increase water prices to keep usage down, the EU is going beyond that to develop a community-wide water pricing policy. The *Communication from the Commission to the Council, European Parliament, and Economic and Social Committee regarding Pricing and Sustainable Management of Water Resources* describes the importance of water pricing policies in relation to sustainability. The communication describes water management as "one of the European Commission's environmental priorities."[57] Pricing and taxation are to be used as policy devices to induce consumers to use water in a more sustainable manner. Water pricing is defined as "the unit or overall amount paid by users for all of the services that they receive in terms of water, including the environment, e.g., wastewater treatment."[58] Thus, the EU has extended water usage to include its use as an environmental sink, as well as an environmental source. The United States has not addressed long-term water usage to this extent.

Table 5.2.
Principles Underlying the EU's Community Water Policy

The following principles are those on which Community Water Policy is based.

- Providing a high level of protection
- The precautionary principle, preventive action
- Damage to be rectified at source
- Principle of polluter pays
- Integration of this policy in other Community policies
- Use of available scientific and technical data
- Variability of environmental conditions in the regions of the Community
- Costs/benefits
- Economic and social development of the community
- International cooperation
- Subsidiarity

Source: European Commission, "Community Water Policy." Accessed 11 January 2003. Available from http://europa.eu.int/scadplus/leg/en/lub/128002a.htm.

LAND-USE PLANNING

In the United States, the freedom to do as one wants with his or her property is considered sacrosanct so that little comprehensive land-use planning has been undertaken in the metropolitan areas. As a result most metropolitan areas suffer from urban sprawl and heavy traffic congestion.

Local governments in the United States generally develop "comprehensive plans" designed to act as blueprints for land use and development in the future. These plans include zoning and subdivision regulations and are revised every five to ten years. However, in some states, large projects that have an areawide impact, such as the location of a power plant or a prison, have been managed under state or regional boards.

As in the United States, most of the EU's metropolitan areas also suffer from traffic congestion. However, in the EU, efforts are being made to specifically address areas of the urban environment, including biodiversity, mobility patterns, and lifestyles. The European Environment Agency's *Dobris Assessment* provided a framework for environmental assessment in an urban environment.[59] The year 1996 was set as a target date for local authorities to have initiated a consultative process for developing Local Agenda 21. (Agenda 21, developed at the United Nations Conference on Environment and Development, held in Rio de Janeiro, 1992, provides a plan of action to be taken in areas in which humans impact the environment.)[60] Thus, many cities in the EU have been moving toward the goal of "educating, mobilizing and responding to the public to promote sustainable development."[61] Many member states have moved to establish national campaigns, such as, Sweden and the United Kingdom. For example, over 70 percent of local authorities in the UK have committed themselves to the Local Agenda 21 process.[62] Cities such as Amsterdam, Berlin, and Copenhagen are developing integrated policies aimed at environmental enhancement. These urban plans consider the use of natural resources and open spaces, efficiency in managing urban flows, the maintenance of cultural and social diversity, and ensuring equal access to resources and services.[63]

While a small number of cities in the United States have developed urban plans and sustainability indicators, these are few, and there is no national impetus to change that. Those that have determined to look toward long-term sustainability, such as Seattle, Washington, often use such standard indicators of environmental quality as air quality, soil erosion, wetlands, and biodiversity. However, Seattle and other cities like it are also addressing long-term sustainability by considering quality-of-life factors, such as the ethnic diversity of teachers, emergency room use, arts instruction, voter participation, and gardening activity.[64] Perhaps U.S. cities such as Seattle will inspire others. In the meantime, the EU is taking major steps toward the goal of sustainability.

In general urban planning (both for infrastructure development and environmental management) has been more long term and comprehensive in the EU than in the United States. We feel this is a function of political culture (particularly lassiez-fare individualism and antigovernmentalism common in parts of the United States, in contrast to greater communitarian acceptance of the role of government that is more common in the EU) and the fact that European urban areas are often much older, and more densely populated and have local planning schemes that have become part of the urban culture.

CONCLUSION

As we have seen in this chapter, the institutional structure of environmental regimes affects the approaches used and the long-term possibilities for environmental improvement. Behind those institutions, the foundations of the structure are built upon the viewpoints and beliefs of the decision makers and cultural norms in their societies. For example, as we have seen, the Bush administration in the United States does not accept that climate change can be averted through the structure of the Kyoto Protocol; EU decision makers believe that this approach can be successful.

It would be tempting to conclude that the reluctance of the George W. Bush administration to embrace the Kyoto Protocol is due to the importance of the U.S. energy industry (notably oil) both to the United States and to the electoral success of the Bush administration. Although these factors undoubtedly come into play, a more persuasive argument can be made, in our opinion, that the relationship between government authority and corporate freedom is a more important variable. Acceptance of the role of government as a guardian of the common good is a greater part of the common political culture in the EU than it is in the United States. This manifests itself in local governments that do comprehensive urban planning in the EU as contrasted to local governments in the United States that struggle over private property rights and the ability to engage in restrictive zoning (something that has been common in the EU for a long time).

In the case of environmental liability, the United States has a well-developed system for addressing environmental hazards. However, the U.S. approach focuses on substances, rather than on the activities that result in contamination by those substances, as does the approach of the EU. By focusing on harms that can result from certain activities, the proposed EU structure has the potential to avert contamination by preventing the harm-causing activities. Much of the difference between the United States and the EU in this area can be traced to the acceptance of the precautionary principle in environmental management. The EU has made the precautionary principle fundamental in environmental policy formation. In the United States policy makers often view the precautionary principle as overly cautious and an infringement on the rights of corporations to act as they wish in the open market. Hence, here too, we feel some fundamental differences in environmental regulation in the United States and the EU can be attributed to differences in political culture and the appropriate role of government in society.

As with environmental liability, the EU does not have the long environmental regulatory history that the United States does, and thus much of the EU legislation appears to be at a more developmental stage. However, unlike the United States, which emphasized economic security and growth, as we witnessed in our discussion of global warming, the EU is more focused on sustainability. In the EU, the economy is one of many factors that are considered in developing legislation, rather than being the primary factor, as it is in much of U.S. legislative decision making. Because of the longer environmental policy history of the United States, much of U.S. legislation has been reactionary, rather than proactive, as it is in the EU. We suspect that over time the EU environmental policy formation system will produce "better" environmental policies (in terms of comprehensiveness, integration of media and NGO participation) than the system in place in the United States. (Ironically to date much of the environmental policy adopted in European countries has borrowed heavily from the United States. We predict that trend will reverse in the future.) We would also anticipate that, because of its orientation toward sustainability, in the long run the EU will evolve into a community that has been able to provide a sustainable and healthy environment, while still providing comfortable living standards for its citizens.

Our final chapter will summarize what we have learned through our examination of the EU. We will extend our understandings of the EU regime to discover what implications this has for the development and improvement of other environmental regimes.

CHAPTER 6

CONCLUSIONS

INTRODUCTION

In our final chapter, we summarize what we have discovered about factors that have supported the efforts of the EU to be successful in environmental protection. We then apply to international environmental regimes what we have learned regarding the policy responses of the EU to environmental challenges and successes of the European Union.

FACTORS LEADING TO ENVIRONMENTAL POLICY SUCCESS IN THE EUROPEAN UNION

What factors have allowed the EU to be successful in addressing environmental policy issues? There are several factors that facilitate the EU's efforts in the environmental arena.

First, the small number of member states may make it easier for representatives to come to agreement on issues. It is much easier to coordinate a small group of countries, especially those who are geographically and culturally similar, than a more diverse group of nations, as is often the case with international environmental regimes.

Second, the EU has placed environmental concerns in a key position in policy making. We have seen that the EU has made environmental protection and sustainability a part of its core objectives. This focus has been a major factor supporting environmental policy success in the EU. The EU has included the goal of sustainability in its treaties and has ruled that environmental protection be included in all other policies developed for the union. The placement of the goal of sustainability in a prominent position both in EU treaties and in EU policy making indicates a willingness and motivation on the part of the policy makers, guided by the public, to locate environmental protection as a basic aim in EU policy. Environmental issues have taken a prominent position in the consciousness of EU citizens and EU policy makers; the citizens of the EU are given a number of avenues through which they can provide input on policy making.

Third, the relatively long history of economic integration among the members of the EU also provides a strong foundation on which to build environmental integration. However, as we have seen, economic objectives are often in conflict with environmental objectives. Yet the EU is striving to integrate those two divergent motivations through the internalization of the environmental costs that are not captured by the price system. Tax credits for the use of alternative fuels and wind power, high taxes on inefficient vehicles, and funding for pesticide reduction programs are all methods that take advantage of the economic motivations of people.

Thus, part of the success of the EU may be due to its large degree of integration in all areas, not just in environmental issues, but also in other areas such as trade and currency. This strong level of integration among the member states brings the representatives together in various contexts, increasing understanding of the positions of the various member states; this, perhaps, makes the representative more conducive to compromise and cooperation.

Next, we examine the structure and approach of the EU as a regime and extend our analysis to other international environmental regimes.

APPLICATION TO INTERNATIONAL ENVIRONMENTAL REGIMES

Many of the challenges facing the European Union's efforts to affect environmental improvements are equivalent to those facing other international regimes, especially those associated with environmental issues. Effective regimes include rules for participation, coordination, conflict resolution, monitoring, and sanctioning.[1] The major obstacles faced by international environmental regimes include funding, implementation, and monitoring. We now review how the EU has dealt with these challenges and extend our findings to international environmental regimes.

The reader will recall that we have defined regimes as social institutions that develop mutually determined rules and decision-making procedures to govern actors.[2] The European Union thus can be considered a regime in the many issue areas in which it has been concerned, including the environment. The EU has, through mutual agreements among the member states, made a series of strategic decisions to place the environment and environmental improvement in a key position in EU treaties and policy making. We will examine how the activities that the EU has pursued follow the same developmental steps, as do other international regimes. We point out the specific factors in the EU that have helped it to become successful in pursuing environmental improvement.

Before the formation of a regime has even begun, it must be recognized that a problem exists, specifically; one that crosses national borders exists, necessitating the involvement of more than one nation. The nations involved must also acknowledge that cooperative effort is necessary to protect and improve the environmental resource in question.[3] Generally, regimes are formed when the actors involved determine that the problems they are facing cannot be solved by individual action but require joint effort. The factors that often lead to the creation of regimes—transboundary problems and the potentially severe outcome associated with the problem, particularly those associated with water and air—have been officially recognized by the EU and its member states. Although it may seem somewhat clear in many cases that the nations must acknowledge the existence of the problem, nations and their leaders do not always share similar perceptions of environmental issues and the need to address these issues. As we have seen in chapter 4, the United States does not recognize a potentially severe problem concerning air pollution; thus the government of the United States has chosen not to become a signatory to the Kyoto Protocol. Thus, nations must be aware of the problem and the need for cooperative solutions if a successful regime is to be initiated.

Besides an awareness and agreement that a cooperative effort in the form of a regime is needed to combat an environmental problem, structures must be developed to provide an avenue for dealing with disagreement and conflict among the signatories to the regime. The EU, as we have seen, has a well-developed system for participation, coordination, and conflict resolution. In chapter 2 we examined the institutional arrangements of the EU. As we saw, the EU provides for representation and the development of coordinated efforts, as well as conflict resolution, among the member states through Parliament and through the European Commission. We also found that the EU has a number of avenues for soliciting public participation, both from individual citizens and from groups of citizens, for example, through the European Economic and Social Committee and through the Committee of the Regions. As we saw, the institutional structure of the EU also provides methods for resolving conflict so that cooperative efforts may succeed.

As we have seen, the EU has been quite successful in developing policies geared toward environmental improvement. The EU, however, does face a problem that often tests the ability of many international environmental regimes to be effective, that of ensuring that policies are appropriately implemented. Ensuring that regime policies are implemented properly requires monitoring the members of the regime to ensure that they are following the rules that the regime has articulated for its signatories. There are many methods that can be utilized for monitoring activities by signatories to the regime. One approach is simply to fund an agency that is given the responsibility of monitoring activity within the regime. The EU does employ agents for monitoring activity among the member states regarding implementation of policy at the national level to ensure that it meets with EU requirements. Another approach to monitoring is through voluntary action. For example, when we examined the case of wind energy development in Ireland in chapter 4, we found that the environmental ministers of the member states had voluntarily joined together to develop approaches to evaluating implementation of environmental directives in the member states. A third alternative to monitoring is to rely on the guardianship activities of the public and nongovernmental organizations (NGOs) to ensure that regime rules are carried out. As we saw in chapter 2, citizens and NGOs in the EU have avenues available to them to seek redress when they believe that environmental policies have not been implemented according to EU directives. Citizens may submit their concerns to the EU ombudsman, or may work through the Court of Justice.

Problems regarding implementation, as well as other environmental problems, particularly those concerning issues between member states, between member states and the EU Commission, or between member states and members of the public may be carried out through the EU Court of Justice system. We have seen that, at times, particular member states who have not met the requirements of EU environmental directives have even had suits filed against them in the Court of Justice (see chapter 2). This is one remedy that is not generally available to international environmental regimes. Often, in international regimes, there are no sanctioning methods to punish regime members who do not abide by the rules of the regime. This, of course, makes it difficult to force compliance; in these cases, the regime members must be willing to voluntarily comply.

Yet it is not enough for the members of the regime to agree to work together on a common problem. Arrangements must be designed to ensure that the regime is effective. We measure the effectiveness of a regime by its ability to realize an improvement in the environment in line with the goals of the regime. Based on our definition of effectiveness, the EU has been a relatively successful regime. We have examined some of the actions taken by the EU that have resulted in environmental improvements. EU treaties include environmental protection and the goal of sustainability as a fundamental aim of EU policy making. The EU developed its Environmental Action Programs to focus on environmental issues. We have also seen how the EU has adjusted taxes, funding measures, and policies in many areas to support environmental improvement.

Another important requirement for successful regimes is the development of a system for gathering and evaluating new information on environmental issues. Recall that in chapter 1 we examined the role of cognitivism in regimes. The EU, as we saw, has instituted arrangements for facilitating the gathering and dissemination of information regarding the environment through the European Environment Agency. As we saw in chapter 2, the EEA has greatly improved the data available on environmental matters in the EU and has provided the public and other interested parties access to this information. In addition, we saw that the public has a right to environmental data from the member states and their environmental agencies. Approaches to increasing information available to the public are also evident in the EU's eco-labeling schemes. If international environmental regimes are to be successful over time, they must be able to gather relevant and valid data and integrate new information into their decision-making structure.

Effective regimes may also require mechanisms for funding the costs associated with environmental improvements; financial limitations often hamper functional implementation of environmental agreements. Funding is often a challenge, especially when the environmental situation targeted by the regime requires costly changes in technology or production processes to achieve results. Regimes such as the EU often find that their members have differing abilities to invest in new processes and consumption patterns that support environmental protection. As we have seen, the EU provides funding for development and implementation of technologies associated with environmental improvement. The EU has thus chosen to lower financial barriers by providing funding to those member states and citizens who may not otherwise be able to invest in measures designed to promote environmental improvements. In other international environmental regimes, the wealthier nations have not always been willing to supply funding or technologies to aid the less prosperous nations. However, if the regime is to be successful in reaching its goals, there must be an acceptance of the need for the transfer of funds and technology that are necessary to meet regime goals.

Another problem facing regimes is associated with national economic factors, that is, the concern that developing and imposing stricter environmental regulations may be economically harmful to a nation's domestic economy. However, with appropriate funding support and transfer of necessary technologies, a nation's domestic economy need not be significantly affected. We have seen that the EU has provided a variety of programs to help the less favored member states improve both their domestic economies and their environments.

Thus, through an examination of the European Union, we have seen how a regime can successfully handle a number of problems often encountered in international environmental regimes. In particular, a more expansive integration of various environmental factors may result in increased environmental improvement. This analysis will facilitate regime designers in developing effective environmental regimes.

CONCLUSION

We have described both the successes and the challenges that have faced the European Union in endeavoring to improve its environment and protect its natural resources. While the EU is different from many other international regimes because of the heavily integrated nature of the member states, it still provides us with valuable knowledge regarding the formation of successful international environmental regimes. Perhaps a key to successful regime formation in other regions is to find ways to increase mutual interdependence and shared perceptions on matters not directly related to the purpose of the regime. Globalization and the universalization of an interdependent market economy may, ironically, provide that shared perception in the future and thereby facilitate environmental regime formation.

APPENDIX 1:

Articles 249–256 of the Treaty on European Union: Provisions Governing the Institutions

Treaty Establishing the European Community as Amended by the Treaty of Amsterdam

Part Five

INSTITUTIONS OF THE COMMUNITY

Title I

Provisions Governing the Institutions

Chapter 2

Provisions common to several institutions

Article 249

In order to carry out their task and in accordance with the provisions of this Treaty, the European Parliament acting jointly with the Council, the Council and the Commission shall make regulations and issue directives, take decision, make recommendations or deliver opinions.

A regulation shall have general application. It shall be binding in its entirely and directly applicable in all Member States.

A directive shall be binding, as to the result to be achieved, upon each Member State to which it is addressed, but shall leave to the national authorities the choice of form and methods.

A decision shall be binding in its entirely upon those to whom it is addressed.

Recommendations and opinions shall have no binding force.

Article 250

1. Where, in pursuance of this Treaty, the Council acts on a proposal from the Commission, unanimity shall be required for an act constituting an amendment to that proposal, subject to Article 251(4) and (5).
2. As long as the Council has not acted, the Commission may alter its proposal at any time during the procedures leading to the adoption of a Community act.

Article 251

1. Where reference is made in this Treaty to this Article for the adoption of an act, the following procedure shall apply.
2. The Commission shall submit a proposal to the European Parliament and the Council.

   The Council, acting by a qualified majority after obtaining the opinion of the European Parliament,

   —If it approves all the amendments contained in the European Parliament's opinion, may adopt the proposed act thus amended;

   —If the European Parliament does not propose any amendments, may adopt the proposed act;

   —Shall otherwise adopt a common position and communicate it to the European Parliament. The Council shall inform the European Parliament fully of the reasons which led it to adopt its common position. The Commission shall inform the European Parliament fully of its position.

If, within three months of such communication, the European Parliament:

(a) approves the common position or has not taken a decision, the act in question shall be deemed to have been adopted in accordance with that common position

(b) rejects, by an absolute majority of its component members, the common position, the proposed act shall be deemed not to have been adopted;

(c) proposes amendments to the common position by an absolute majority of its component members, the amended text shall be forwarded to the Council and to the Commission, which shall deliver an opinion on those amendments.

3. If, within three months of the matter being referred to it, the Council, acting by a qualified majority, approves all the amendments of the European Parliament, the act in question shall be deemed to have been adopted in the form of the common position thus amended; however, the Council shall act unanimously on the amendments on which the Commission has delivered a negative opinion. If the Council does not approve all the amendments, the President of the Council, in agreement with the President of the European Parliament, shall within six weeks convene a meeting of the Conciliation Committee.
4. The Conciliation Committee, which shall be composed of the members of the Council or their representatives and an equal number of representatives of the European Parliament, shall have the task of reaching agreement on a joint text, by a qualified majority of the members of the Council or their representatives and by a majority of the representatives of the European Parliament. The Commission shall take part in the Conciliation Committee's proceedings and shall take all the necessary initiatives with a view to reconciling the positions of the European Parliament and the Council. In fulfilling this task, the Conciliation Committee shall address the common position on the basis of the amendments proposed by the European Parliament.
5. If, within six weeks of its being convened, the Conciliation Committee approves a joint text, the European Parliament, acting by an absolute majority of the votes cast, and the Council, acting by a qualified majority, shall each have a period of six weeks from that approval in which to adopt the act in question in accordance with the joint text. If either of the two institutions fails to approve the proposed act within that period, it shall be deemed not to have been adopted.
6. Where the Conciliation Committee does not approve a joint text, the proposed act shall be deemed not to have been adopted.
7. The periods of three months and six weeks referred to in this Article shall be extended by a maximum of one month and two weeks respectively at the initiative of the European Parliament or the Council.

Article 252

Where reference is made in this Treaty to this Article for the adoption of an act, the following procedure shall apply:

(a) The Council, acting by a qualified majority on a proposal from the Commission and after obtaining the opinion of the European Parliament, shall adopt a common position.

(b) The Council's common position shall be communicated to the European Parliament. The Council and the Commission shall inform the European Parliament fully of the reasons which led the Council to adopt its common position and also of the Commission's position.

If, within three months of such communication, the European Parliament approves this common position or has not taken a decision within that period, the Council shall definitively adopt the act in question in accordance with the common position.

(c) The European Parliament may, within the period of three months referred to in point (b), by an absolute majority of its component Members, propose amendments to the Council's common position. The European Parliament may also, by the same majority, reject the Council's common position. The result of the proceedings shall be transmitted to the Council and the Commission.

If the European Parliament has rejected the Council's common position, unanimity shall be required for the Council to act on a second reading.

(d) The Commission shall, within a period of one month, re-examine the proposal on the basis of which the Council adopted its common position, by taking into account the amendments proposed by the European Parliament.

The Commission shall forward to the Council, at the same time as its re-examined proposal, the amendments of the European Parliament which it has not accepted, and shall express its opinion on them. The Council may adopt these amendments unanimously.

(e) The Council, acting by a qualified majority, shall adopt the proposal as re-examined by the Commission.

Unanimity shall be required for the Council to amend the proposal as re-examined by the Commission.

(f) In the cases referred to in points (c), (d), and (e), the Council shall be required to act within a period of three months. If no decision is taken within this period, the Commission proposal shall be deemed not to have been adopted.

(g) The periods referred to in points (b) and (f) may be extended by a maximum of one month by common accord between the Council and the European Parliament.

Article 253

Regulations, directives and decisions adopted jointly by the European Parliament and the Council, and such acts adopted by the Council or the Commission, shall state the reasons on which they are based and shall refer to any proposals or opinions which were required to be obtained pursuant to this Treaty.

Article 254

1. Regulations, directives and decisions adopted in accordance with the procedure referred to in Article 251 shall be signed by the President of the European Parliament and by the President of the Council and published in the Official Journal of the European Communities. They shall enter into force on the date specified in them or, in the absence thereof, on the twentieth day following that of their publication.
2. Regulations of the Council and of the Commission, as well as directives of these institutions which are addressed to all Member States, shall be published in the Official Journal of the European Communities. They shall enter into force on the date specified in them or, in the absence thereof, on the twentieth day following that of their publication.
3. Other directives, and decisions, shall be notified to those to whom they are addressed and shall take effect upon such notification.

Article 255

1. Any citizen of the Union, and any natural or legal person residing or having its registered office in a Member States, shall have a right of access to European Parliament, Council and Commission documents, subject to the principles and the conditions to be defined in accordance with paragraphs 2 and 3.
2. General principles and limits on grounds of public or private interest governing this right of access to documents shall be determined by the Council, acting in accordance with the procedure referred to in Article 251 within two years of the entry into force of the Treaty of Amsterdam.

3. Each institution referred to above shall elaborate in its own Rules of Procedure specific provisions regarding access to its documents.

Article 256

Decisions of the Council or of the Commission which impose a pecuniary obligation on persons other than States, shall be enforceable.

Enforcement shall be governed by the rules of civil procedure in force in the State in the territory of which it is carried out. The order for its enforcement shall be appended to the decision, without other formality than verification of the authenticity of the decision, by the national authority which the government of each Member State shall designate for this purpose and shall make known to the Commission and to the Court of Justice.

When these formalities have been completed on application by the party concerned, the latter may proceed to enforcement in accordance with the national law, by bringing the matter directly before the competent authority.

Enforcement may be suspended only by a decision of the Court of Justice. However, the courts of the country concerned shall have jurisdiction over complaints that enforcement is being carried out in an irregular manner.

APPENDIX 2:

Articles 174–176 of the Treaty on European Union: Environment

Treaty Establishing the European Community as Amended by the Treaty of Amsterdam

Part Five

INSTITUTIONS OF THE COMMUNITY

Title XIX

ENVIRONMENT

Article 174

1. Community policy on the environment shall contribute to pursuit of the following objectives:[1]

—Preserving, protecting and improving the quality of the environment;

—Protecting human health;

—Prudent and rational utilization of natural resources;

—Promoting measures at international level to deal with regional or worldwide environmental problems.

2. Community policy on the environment shall aim at a high level of protection taking into account the diversity of situations in the various regions of the Community. It shall be based on the precautionary principle and on the principles that preventive action should be taken, that environmental damage should as a priority be rectified at source and that the polluter should pay.

   In this context, harmonization measures answering environmental protection requirements shall include, where appropriate, a safeguard clause allowing Member States to take provisional measures, for non-economic environmental reasons, subject to a Community inspection procedure.

3. In preparing its policy on the environment, the Community shall take account of:

—Available scientific and technical data;

—Environmental conditions in the various regions of the Community;

—The potential benefits and costs of action or lack of action;

—The economic and social development of the Community as a whole and the balanced development of its regions.

4. Within their respective spheres of competence, the Community and the Member States shall cooperate with third countries and with the competent international organizations. The arrangements for Community cooperation may be the subject of agreements between the Community and the third parties concerned, which shall be negotiated and concluded in accordance with Article 300.

The previous subparagraph shall be without prejudice to Member States' competence to negotiate in international bodies and to conclude international agreements.

Article 175

1. The Council, acting in accordance with the procedure referred to in Article 251 and after consulting the Economic and Social Committee and the Committee of the Regions, shall decide what action is to be taken by the Community in order to achieve the objectives referred to in Article 174.

2. By way of derogation from the decision-making procedure provided for in paragraph 1 and without prejudice to Article 95,[2] the Council, acting unanimously on a proposal from the Commission and after consulting the European Parliament, the Economic and Social Committee, and the Committee of the Regions, shall adopt:

—Provisions primarily of a fiscal nature;

—Measures concerning town and country planning, land use with exception of waste management and measures of a general nature, and management of water resources;

—Measures significantly affecting a Member State's choice between different energy sources and the general structure of its energy supply.

The Council may, under the conditions laid down in the preceding subparagraph, define those matters referred to in this paragraph on which decisions are to be taken by a qualified majority.

3. In other areas, general action programmes setting out priority objective to be attained shall be adopted by the Council, acting in accordance with the procedure referred to in Article 251 and after consulting the Economic and Social Committee and the Committee of the Regions.

   The Council, acting under the terms of paragraph 1 or paragraph 2 according to the case, shall adopt the measures necessary for the implementation of these programmes.

4. Without prejudice to certain measure of a Community nature, the Member States shall finance and implement the environment policy.

5. Without prejudice to the principle that the polluter should pay, if a measure based on the provisions of paragraph 1 involves costs deemed disproportionate for the public authorities of a Member State, the Council shall, in the act adopting that measure, lay down appropriate provisions in the form of:

—Temporary derogations, and/or

—Financial support from the Cohesion Fund set up pursuant to Article 161.

Article 176

The protective measures adopted pursuant to Article 175 shall not prevent any Member State from maintaining or introducing more stringent protective measures. Such measures must be compatible with this Treaty. They shall be notified to the Commission.

APPENDIX 3:

The Sixth Community Environment Action Program

Decision No 1600/2002/EC of the EUROPEAN PARLIAMENT and of THE COUNCIL

Of 22 July 2002

Laying down the Sixth Community Environment Action Programme

THE EUROPEAN PARLIAMENT AND THE COUNCIL OF THE EUOR-PEAN UNION, Having regard to the Treaty establishing the European Community, and in particular Article 175(3) thereof,[1]

Having regard to the proposal from the Commission[2] Having regard to the opinion of the Economic and Social Committee,[3]

Having regard to the opinion of the Committee of the Regions[4]

Acting in accordance with the procedure laid down in Article 251 of the Treaty,[5] in the light of the joint text approved by the Conciliation Committee on 1 May 2002,

Whereas:

(1) A clean and healthy environment is essential for the well-being and prosperity of society, yet continued growth at a global level will lead to continuing pressures on the environment.

(2) The Community's fifth environmental action programme 'Towards Sustainability' ended on 31 December 2002 having delivered a number of important improvements.

(3) Continued effort is required in order to meet the environmental objectives and targets already established by the Community and there is a need for the Sixth Environmental Action Programme (the 'Programme') set out in this Decision.

(4) A number of serious environmental problems persist and new ones are emerging which require further action.

(5) Greater focus on prevention and the implementation of the precautionary principle is required in developing an approach to protect human health and the environment.

(6) A prudent use of natural resources and the protection of the global eco-system together with economic prosperity and a balanced social development are a condition for sustainable development.

(7) The Programme aims at a high level of protection of the environment and human health and at a general improvement in the environment and quality of life, indicates priorities for the environmental dimension of the Sustainable Development Strategy and should be taken into account when bringing forward actions under the Strategy.

(8) The Programme aims to achieve a decoupling between environmental pressures and economic growth whilst being consistent with the principle of subsidiarity and respecting the diversity of conditions across the various regions of the European Union.

(9) The Programme established environmental priorities for a Community response focusing in particular on climate change, nature and biodiversity, environment and health and quality of life, and natural resources and wastes.

(10) For each of these areas key objectives and certain targets are indicated and a number of actions are identified with a view to achieving the said targets. These objectives and targets constitute performance levels or achievements to be aimed at.

(11) The objectives, priorities and actions of the Programme should contribute to sustainable development in the candidate countries and endeavor to ensure the protection of the natural assets of theses communities.

(12) Legislation remains central to meeting environmental challenges and full and correct implementation of the existing legislation is a priority. Other options for achieving environmental objectives should also be considered.

(13) The Programme should promote the process of integration of environmental concerns in to all Community polities and activities in line with Article 6 of the Treaty in order to reduce the pressures on the environment from various sources.

(14) A strategic integrated approach, incorporating new ways of working with the market, involving citizens, enterprises and other stakeholders is needed in order to induce necessary changes in both production and public and private consumption patterns that

influence negatively the state of, and trends in, the environment. This approach should encourage sustainable use and management of land and sea.

(15) Provision for access to environmental information and to justice and for public participation in policy-making will be important to the success of the Programme.

(16) Thematic strategies will consider the range of options and instruments required for dealing with a series of complex issues that require a broad and multi-dimensional approach and will propose the necessary actions, involving where appropriate the European Parliament and the Council.

(17) There is scientific consensus that human activity is causing increases in concentrations of green house gases, leading to higher global temperatures and disruption to the climate.

(18) The implications of climate change for human society and for nature are severe and necessitate mitigation. Measures to reduce emissions of greenhouse gases can be implemented without a reduction in levels of growth and prosperity.

(19) Regardless of the success of mitigation, society needs to adapt to and prepare for the effects of climate change.

(20) Healthy and balanced natural systems are essential for supporting life on the planet.

(21) There is considerable pressure from human activity on nature and biodiversity. Action is necessary to counteract pressures arising notably from pollution, the introduction of non-native species, potential risks from releasing genetically modified organisms and the way in which the land and sea are exploited.

(22) Soil is a finite resource that is under environmental pressure.

(23) Despite improvements in environmental standards, there is increased likelihood of a link between environmental degradation and certain human illnesses. Therefore the potential risks arising, for example, from emissions and hazardous chemicals, pesticides, and from noise should be addressed.

(24) Greater knowledge is required on the potential negative impacts arising from the use of chemicals and the responsibility for generating knowledge should be placed on producers, importers and downstream users.

(25) Chemicals that are dangerous should be replaced by safer chemicals or safer alternative technologies not entailing the use of chemicals, with the aim of reducing risks to man and the environment.

(26) Pesticides should be used in a sustainable way so as to minimize negative impacts for human health and the environment.

(27) The urban environment is home to some 70% of the population and concerted efforts are needed to ensure a better environment and quality of life in towns and cities.

(28) There is a limited capacity of the planet to meet the increasing demand for resources and to absorb the emissions and waste resulting from their use and there is evidence that the existing demand exceeds the carrying capacity of the environment in several cases.

(29) Waste volumes in the Community continue to rise, a significant quantity of these being hazardous, leading to loss of resources and to increased pollution risks.

(30) Economic globalization means that environmental action is increasingly needed at international level, including on transport policies, requiring new responses from the Community linked to policy related to trade, development and external affairs enabling sustainable development to be pursued in other countries. Good governance should make a contribution to this end.

(31) Trade, international investment flows and export credits should make a more positive contribution to the pursuit of environmental protection and sustainable development.

(32) Environmental policy-making, given the complexities of the issues, needs to be based on best available scientific and economic assessment, and on knowledge of the state and trends of the environment, in line with Article 174 of the Treaty.

(33) Information to policy makers, stakeholders and the general public has to be relevant, transparent, up to date and easily understandable.

(34) Progress towards meeting environmental objectives needs to be measured and evaluated.

(35) On the basis of an assessment of the state of the environment, taking account of the regular information provided by the

European Environment Agency, a review of progress and an assessment of the need to change orientation should be made at the mid term point of the Programme,

HAVE DECIDED AS FOLLOWS:

Article 1

Scope of the Programme

1. This Decision establishes a programme of Community action on the environment (hereinafter referred to as the 'Programme'). It addresses the key environmental objectives and priorities based on an assessment of the state of the environment and of prevailing trends including emerging issues that require a lead from the Community. The Programme should promote the integration of environmental concerns in all Community policies and contribute to the achievement of sustainable development throughout the current and future enlarged Community. The Programme furthermore provides for continuous efforts to achieve environmental objectives and targets already established by the Community.

2. The Programme sets out the key environmental objectives to be attained. It establishes, where appropriate, targets and timetables. The objectives and targets should be fulfilled before expiry of the Programme, unless otherwise specified.

3. The Programme shall cover a period of ten years starting from 22 July 2002. Appropriate initiatives in the different policy areas with the aim of meeting the objectives shall consist of a range of measures including legislation and the strategic approaches outlined in Article 3. These initiatives should be presented progressively and at the latest by four years after the adoption of this Decision.

4. The objectives respond to the key environmental priorities to be met by the Community in the following areas:

—climate change,

—nature and biodiversity,

—environment and health and quality of life,

—natural resources and wastes.

Article 2

Principles and overall aims

The Programme constitutes a framework for the Community's environmental policy during the period of the Programme with the aim of ensuring a high level of protection, taking into account the principle of subsidiarity and the diversity of situations in the various regions of the Community, and of achieving a decoupling between environmental pressures and economic growth. It shall be based particularly on the polluter-pays principle, the precautionary principle and preventive action, and the principle of rectification of pollution at the source.

The Programme shall form a basis for the environmental dimension of the European Sustainable Development Strategy and contribute to the integration of environmental concerns into all Community policies, *inter alia* by setting out environmental priorities for the Strategy.

1. The Programme aims at:

—emphasizing climate change as an outstanding challenge of the next 10 years and beyond and contributing to the long term objective of stabilizing greenhouse gas concentrations in the atmosphere at a level that would prevent dangerous anthropogenic interference with the climate system. Thus a long term objective of a maximum global temperature increase of 2° Celsius over pre-industrial levels and a CO_2 concentration below 550 ppm shall guide the Programme. In the longer term this is likely to require a global reduction in emissions of greenhouse gases by 70% as compared to 990 as identified by the Intergovernmental Panel on Climate Change (IPCC);

—protecting, conserving, restoring and developing the functioning of natural systems, natural habitats, wild flora and fauna with the aim of halting desertification and the loss of biodiversity, including diversity of genetic resources, both in the European Union and on a global scale;

—contributing to a high level of quality of life and social well being for citizens by providing an environment where the level of pollution does not give rise to harmful effects on human health and the environment and by encouraging a sustainable urban development;

—better resource efficiency and resource and waste management to bring about more sustainable production and consumption patterns, thereby decoupling the use of resources and the generation of waste from the rate of economic growth and aiming to ensure that the consumption of renewable and non-renewable resources does not exceed the carrying capacity of the environment.

2. The Programme shall ensure that environmental objectives, which should focus on the environmental outcomes to be achieved, are met by the most effective and appropriate means available, in the light of the principles set out in paragraph 1 and the strategic approaches set out in Article 3. Full consideration shall be given to ensuring that the Community's environmental policy-making is undertaken in an integrated way and to all available options and instruments, taking into account regional and local differences, as well as ecologically sensitive areas, with an emphasis on:

—developing European initiatives to raise the awareness of citizens and local authorities;

—extensive dialogue with stakeholders, raising environmental awareness and public participation;

—analysis of benefits and costs, taking into account the need to internalize environmental costs;

—the best available scientific evidence, and the further improvement of scientific knowledge through research and technological development;

—data and information on the state and trends of the environment.

3. The Programme shall promote the full integration of environmental protection requirements into all Community policies and actions by establishing environmental objectives and, where appropriate, targets and timetables to be taken into account in relevant policy areas.

Furthermore, measures proposed and adopted in favor of the environment should be coherent with the objectives of the economic and social dimensions of sustainable development and vice versa.

4. The Programme shall promote the adoption of policies and approaches that contribute to the achievement of sustainable development in the countries which are candidates for accession ('Candidate Countries') building on the transposition and implementation of the acquis. The enlargement process should sustain and protect the environmental assets of the Candidate Countries such as wealth of biodiversity, ad should maintain and strengthen sustainable production and consumption and land use patterns and environmentally sound transport structures through:

—integration of environmental protection requirements into Community Programmes including those related to development of infrastructure;

—promotion of transfer of clean technologies to the Candidate Countries;

—extended dialogue and exchange of experience with the national and local administrations in the Candidate Countries on sustainable development and preservation of their environmental assets;

—cooperation with civil society, environmental non-governmental organizations (NGOs) and business in the Candidate Countries to help raise public awareness and participation;

—encouraging international financing institutions and the private sector to support the implementation of and compliance with the environmental acquis in the Candidate Countries and to pay due attention to integrating environmental concerns into the activities of the economic sector.

5. The Programme shall stimulate:

—the positive and constructive role of the European Union as a leading partner in the protection of the global environment and in the pursuit of a sustainable development;

—the development of a global partnership for environment and sustainable development;

—the integration of environmental concerns and objectives into all aspects of the Community's external relations.

Article 3

Strategic approaches to meeting environmental objectives

The aims and objectives set out in the Programme shall be pursued, *inter alia*, by the following means:

1. Development of new Community legislation and amendment of existing legislation, where appropriate;
2. Encouraging more effective implementation and enforcement of Community legislation on the environment and without prejudice to the Commission's right to initiate infringement proceedings. This requires:

—increased measures to improve respect for Community rules on the protection of the environment and addressing infringements of environmental legislation;

—promotion of improved standards of permitting, inspection, monitoring and enforcement by Member States;

—a more systematic review of the application of environmental legislation across the Member States;

—improved exchange of information on best practice on implementation including by the European Network for the Implementation and enforcement of Environmental Law (IMPEL network) within the framework of its competencies;

3. Further efforts for integration of environmental protection requirements into the preparation, definition and implementation of Community policies and activities in the different policy areas are needed. Further efforts are necessary in different sectors including consideration of their specific environmental objectives, targets, timetables and indicators. This requires:

—ensuring that the integration strategies produced by the Council in different policy areas are translated into effective action and contribute to the implementation of the environmental aims and objectives of the Programme;

—consideration, prior to their adoption, of whether action in the economic and social fields, contribute to and are coherent with the objectives, targets and time frame of the Programme;

—establishing appropriate regular internal mechanisms in the Community institutions, taking full account of the need to promote transparency and access to information, to ensure that environmental considerations are fully reflected in Commission policy initiatives, including relevant decisions and legislative proposals;

—regular monitoring, via relevant indicators, elaborated where possible on the basis of a common methodology for each sector, and reporting on the process of sectoral integration;

—further integration of environmental criteria into Community funding programmes without prejudice to existing ones;

—full and effective use and implementation of Environmental Impact Assessment and Strategic Environmental Assessment;

—that the objectives of the Programme should be taken into account in future financial perspective reviews of Community financial instruments;

4. Promotion of sustainable production and consumption patterns by effective implementation of the principles set out in Article 2, to internalize the negative as well as the positive impacts on the environment through the use of a blend of instruments, including market based and economic instruments. This requires, *inter alia*:

—encouraging reforms of subsidies that have considerable negative effects on the environment and are incompatible with sustainable development, *inter alia* by establishing, by the mid-term review, a list of criteria allowing such environmentally negative subsidies to be recorded, with a view to gradually eliminating them;

—analyzing the environmental efficiency of tradeable environmental permits as a generic instrument and of emission trading with a view to promoting and implementing their use where feasible;

—promoting and encouraging the use of fiscal measures such as environmentally related taxes and incentives, at the appropriate national or Community level;

—promoting the integration of environmental protection requirements in standardization activities;

5. Improving collaboration and partnership with enterprises and their representative bodies and involving the social partners, consumers and their organizations, as appropriate, with a view to improving the environmental performance of enterprises and aiming at sustainable production patterns. This requires:

—promoting an integrated product policy approach throughout the Programme that will encourage the taking into account of environmental requirements throughout the life-cycle of products, and more wide-spread application of environmentally friendly processes and products;

—encouraging wider uptake of the Community's Eco-Management and Audit Scheme (EMAS)[6] and developing initiatives to encourage companies to publish rigorous and independently verified environmental or sustainable development performance reports;

—establishing a compliance assistance programme, with specific help for small and medium enterprises;

—stimulating the introduction of company environmental performance award schemes;

—stimulating product innovation with the aim of greening the market including through improved dissemination of results of the LIFE Programme[7]

—encouraging voluntary commitments or agreements to achieve clear environmental objectives, including setting out procedures in the event of non-compliance;

6. To help ensure that individual consumers, enterprises and public bodies in their roles as purchasers, are better informed about the processes and products in terms of their environmental impact with a view to achieving sustainable consumption patterns. This requires:

—encouraging the uptake of eco-labels and other forms of environmental information and labeling that allow consumers to compare environmental performance between products of the same type;

—encouraging the use of reliable self-declared environmental claims and preventing misleading claims;

—promoting a green public procurement policy, allowing environmental characteristics to be taken into account and the possible integration of environmental life cycle, including the production phase, concerns in the procurement procedures while respecting Community competition rules and the internal market, with guidelines on best practice and starting a review of green procurement in Community Institutions;

6. To support environmental integration in the financial sector. This requires:

—considering a voluntary initiative with the financial sector, covering guide-lines for the incorporation of data on environmental cost in company annual financial reports, and the exchange of best policy practices between Member States;

—calling on the European Investment Bank to strengthen the integration of environmental objectives and considerations into its lending activities in particular with a view to supporting a sustainable development of Candidate Countries;

—promoting integration of environmental objectives and considerations into the activities of other financial institutions such as the European Bank for reconstruction and Development;

7. To create a Community liability regime required *inter alia*:

—legislation on environmental liability;

8. To improve collaboration and partnership with consumer groups and NGOs and promote better understanding of and participation in environmental issues amongst European citizens requires:

—ensuring access to information, participation and justice through early ratification of the Aarhus Convention[8] by the Community and by Member States;

—supporting the provision of accessible information to citizens on the state and trends of the environment in relation to social, economic and health trends;

—general raising of environmental awareness;

—developing general rules and principles for good environmental governance in dialogue processes;

9. To encourage and promote effective and sustainable use and management of land and sea taking account of environmental concerns. This requires, while fully respecting the subsidiarity principle, the following:

—promoting best practice with respect to sustainable land use planning, which takes account of specific regional circumstances with particular emphasis on the Integrated Coastal Zone Management programme;

—promoting best practices and supporting networks fostering the exchange of experience on sustainable development including urban areas, sea, coastline, mountain areas, wetlands and other areas of a sensitive nature;

—enhancing the use, increasing resources and giving broader scope for agri-environment measures under the Common Agricultural Policy;

—encouraging Member States to consider using regional planning as an instrument for improving environmental protection for the citizen and promoting the exchange of experience on sustainable regional development, particularly in urban and densely populated areas.

Article 4

Thematic strategies

1. Actions in Articles 5 to 8 shall include the development of thematic strategies and the evaluation of existing strategies for priority environmental problems requiring a broad approach. These strategies should include an identification of the proposals that are required to reach the objectives set out in the Programme and the procedures foreseen for their adoption. These strategies shall be submitted to the European Parliament and Council and shall, where appropriate take the form of a Decision of the European Parliament and of the Council to be adopted in accordance with the procedure laid down in Article 251 of the treaty. Subject to the legal base of the proposal, the legislative proposals arising from these strategies shall be adopted in accordance with the procedure laid down in Article 251 of the Treaty.

2. The thematic strategies may include approaches among those outlines in Article 3 and in Article 9 and relevant qualitative and quantitative environmental targets and timetables against which the measures foreseen can be measured and evaluated.

3. The thematic strategies should be developed and implemented in close consultation with the relevant parties, such as NGOs, industry, other social partners and public authorities, while ensuring, as appropriate, consultation of Candidate Countries in this process.

4. The thematic strategies should be presented to the European Parliament and the Council within 3 years of the adoption of the Programme at the latest. The mid-term report, in which the Commission evaluates the progress made in implementing the Programme, shall include a review of the thematic strategies.

5. The Commission shall report annually to the European Parliament and the Council on the progress in the development and implementation of the strategies and on their effectiveness.

Article 5

Objectives and priority areas for action on tackling climate change

1. The aims set out in Article 2 should be pursued by the following objectives:

—ratification and entering into force of the Kyoto Protocol to the United Nations framework Convention on climate change by 2002 and fulfillment of its commitment of an 8% reduction in emissions by 2008–12 compared to 1990 levels for the European Community as a whole, in accordance with the commitment of each Member State set out in the Council Conclusions of 16 and 17 June 1998;

—realization by 2005 of demonstrable progress in achieving the commitments under the Kyoto Protocol;

—placing the Community in a credible position to advocate an international agreement on more stringent reduction targets for the second commitment period provided for by the Kyoto Protocol. This agreement should aim at cutting emissions significantly, taking full account, *inter alia*, of the findings of the IPCC 3rd Assessment Report, and take into account the necessity to move towards a global equitable distribution of greenhouse gas emissions.

2. These objectives shall be pursued by means, *inter alia*, of the following priority actions:

(i) Implementing international climate commitments including the Kyoto Protocol by means of

(a) Examining the results of the European Climate Change Programme and adopting effective common and coordinated policies and measures on its basis, as appropriate, for various sectors complementary to domestic actions in the Member States;

(b) Working towards the establishment of a Community framework for the development of effective CO_2 emissions trading with the possible extension to other greenhouse gases;

(c) Improving monitoring of greenhouse gases and of progress towards delivering Member States commitments made under the Internal Burden Sharing Agreement;

(ii) Reducing greenhouse gas emissions in the energy sector:

(a) Undertaking as soon as possible an inventory and review of subsidies that counteract an efficient and sustainable use of energy with a view to gradually phasing them out;

(b) Encouraging renewable and lower carbon fossil fuels for power generation;

(c) Encouraging the use of renewable energy sources, including the use of incentives, including at the local level, with a view to meeting the indicative target of 12% of total energy use by 1020;

(d) Introducing incentives to increase Combined Heat and Power and implement measures aiming at doubling the overall share of Combined Heat and Power in the Community as a whole to 18% of the total gross electricity generation;

(e) Prevent and reduce methane emissions from energy production and distribution;

(f) Promoting energy efficiency;

(iii) Reducing greenhouse gas emissions in the transport sector:

(a) Identifying and undertaking specific actions to reduce greenhouse gas emissions from aviation if no such action is agreed within the International Civil Aviation Organization by 2002;

(b) Identifying and undertaking specific actions to reduce greenhouse gas emissions from marine shipping if no such action is agreed within the International Maritime Organization by 2003;

(c) Encouraging a switch to more efficient and cleaner forms of transport including better organization and logistics;

(d) In the context of the EU target of an 8% reduction in greenhouse gas emissions, inviting the Commission to submit by the end of 2002 a Communication on quantified environmental objectives for a sustainable transport system;

(e) Identifying and undertaking further specific action, including any appropriate legislation, to reduce greenhouse gas emissions from motor vehicles including N_2O;

(f) Promoting the development and use of alternative fuels and of low-fuel-consuming vehicles with the aim of substantially and continually increasing their share;

(g) Promoting measures to reflect the full environmental costs in the price of transport;

(h) Decoupling economic growth and the demand for transport with the aim of reducing environmental impacts;

(iv) Reducing greenhouse gas emissions in industrial production:

(a) Promoting eco-efficiency practices and techniques in industry;

(b) Developing means to assist SMEs to adapt, innovate and improve performance;

(c) Encouraging the development of more environmentally sound and technically feasible alternatives, including the establishment of

Community measures, aiming at reducing emissions, phasing out the production where appropriate and feasible and reducing the use of industrial fluorinated gases HFCs (hydrofluorocarbons), PFCs (perfluorocarbons) and SF_6 (sulphur hexafluoride);

(v) Reducing greenhouse gas emissions in other sectors:

(a) Promoting energy efficiency notably for heating, cooling, and hot tap water in the design of buildings;

(b) Taking into account the need to reduce greenhouse gas emissions, alongside with other environmental considerations, in the Common agricultural policy and in the Community's waste management strategy;

(vi) Using other appropriate instruments such as:

(a) promoting the use of fiscal measures, including a timely and appropriate Community framework for energy taxation, to encourage a switch to more efficient energy use, cleaner energy and transport and to encourage technological innovation;

(b) encouraging environmental agreements with industry sectors on greenhouse gas emission reductions;

(c) Ensuring climate change as a major theme of Community policy for research and technological development and for national research programmes.

3. In addition to the mitigation of climate change, the Community should prepare for measures aimed at adaptation to the consequences of climate change, by:

—reviewing Community policies, in particular those relevant to climate change, so that adaptation is addressed adequately in investment decisions;

—encouraging regional climate modeling and assessments both to prepare regional adaptation measures such as water resources management, conservation of biodiversity, desertification and flooding prevention and to support awareness raising among citizens and business.

4. It must be ensured that the climate challenge is taken into account in the Community's enlargement. This will require, *inter alia*, the following actions with Candidate Countries:

—supporting capacity building, for the application of domestic measures for the use of the Kyoto mechanisms and improved reporting and emission monitoring;

—supporting a more sustainable transport and energy sector;

—ensuring that cooperation with candidate countries is further strengthened on climate change issues.

5. Combating climate change will form an integral part of the European Union's external relations policies and will constitute one of the priorities in its sustainable development policy. This will require concerted and coordinated efforts on the part of the Community and its Member States with a view to:

—capacity-building to assist developing countries and countries with economies in transition for example through encouraging projects in connection with the Clean Development Mechanism (CDM) in the Kyoto Protocol and joint implementation;

—responding to identified technology-transfer needs;

—assisting with the challenge of adapting to climate change in the countries concerned.

Article 6

Objectives and priority areas for action on nature and biodiversity

1. The aims set out in Article 2 should be pursued by the following objectives:

—halting biodiversity decline with the aim to reach this objective by 2010, including prevention and mitigation of impacts of invasive alien species and genotypes;

—protection and appropriate restoration of nature and biodiversity from damaging pollution;

—conservation, appropriate restoration and sustainable use of marine environment, coasts and wetlands;

—conservation and appropriate restoration of areas of significant landscape values including cultivated as well as sensitive areas;

—promotion of a sustainable use of the soil, with particular attention to preventing erosion, deterioration, contamination and desertification.

2. These objectives shall be pursued by means of the following priority actions, taking into account the principle of subsidiarity, based on the existing global and regional conventions and strategies and full implementation of the relevant Community acts. The ecosystem approach, as adopted in the Convention on Biological Diversity,[9] should be applied whenever appropriate:

(a) On biodiversity:

—ensuring the implementation and promoting the monitoring and assessment of the Community's biodiversity strategy and the relevant action plans, including through a programme for gathering data and information, developing the appropriate indicators, and promoting the use of best available techniques and of best environmental practices;

—promoting research on biodiversity, genetic resources, ecosystems and interactions with human activities;

—developing measures to enhance sustainable use, sustainable production and sustainable investments in relation to biodiversity;

—encouraging coherent assessment, further research and cooperation on threatened species;

—promoting at the global level a fair and equitable sharing of benefits arising from the use of genetic resources to implement Article 15 of the Convention on Biological Diversity on access to genetic resources originating from third countries;

—developing measures aimed at the prevention and control of invasive alien species including alien genotypes;

—establishing the Natura 2000 network and implementing the necessary technical and financial instruments and measures required for its full implementation and for the protection, outside of the Natura 2000 areas, of species protected under the Habitats and Birds Directives;

—promoting the extension of the Natura 2000 network to the Candidate Countries;

(b) On accidents and disasters:

—promoting Community coordination to actions by Member States in relation to accidents and natural disasters by, for example, setting up a network for exchange of prevention practices and tools;

—developing further measures to help prevent the major accident hazards with special regards to those arising from pipelines, mining, marine transport of hazardous substances and developing measures on mining waste;

(c) A thematic strategy on soil protection, addressing the prevention of, *inter alia*, pollution, erosion, desertification, land degradation, land-take and hydrogeological risks taking into account regional diversity, including specificities of mountain and arid areas;

(d) Promoting sustainable management of extractive industries with a view to reduce their environmental impact;

(e) Promoting the integration of conservation and restoration of the landscape values into other policies including tourism, taking account of relevant international instruments;

(f) Promoting the integration of biodiversity consideration in agricultural policies and encouraging sustainable rural development, multi-functional and sustainable agriculture, through:

—encouraging full use of current opportunities of the Common Agriculture Policy and other policy measures;

—encouraging more environmentally responsible farming, including, where appropriate, extensive production methods, integrated farming practices, organic farming and agro-biodiversity, in future reviews of the Common Agricultural Policy, taking account of the need for a balanced approach to the multifunctional role of rural communities;

(g) Promoting sustainable use of the seas and conservation of marine ecosystems, including sea beds, estuarine and coastal areas, paying special attention to sites holding a high biodiversity value, through:

—promoting greater integration of environmental considerations in the Common Fisheries Policy, taking the opportunity of its review in 2002;

—a thematic strategy for the protection and conservation of the marine environment taking into account, *inter alia*, the terms and implementation obligations of marine Conventions, and the need to reduce emissions and impacts of sea transport and other sea and land-based activities;

—promoting integrated management of coastal zones;

—further promote the protection of marine areas, in particular with the Natura 2000 network as well as by other feasible Community means;

(h) Implementing and further developing strategies and measures on forests in line with the forest strategy for the European Union, taking account the principle of subsidiarity and biodiversity considerations, incorporating the following elements:

—improving existing Community measures which protect forests and implementing sustainable forest management, *inter alia*, through national forest programmes, in connection with rural development plans, with increased emphasis on the monitoring of the multiple roles of forests in line with recommendations adopted by the Ministerial;

—encouraging the effective coordination between all policy sectors involved in forestry, including the private sector, as well as the coordination of all stakeholders involved in forestry issues;

—stimulating the increase of the market share for sustainably produced wood, *inter alia*, through encouraging certification for sustainable forest management and encouraging labeling of related products;

—continuing the active participation of the Community and of Member States in the implementation of global and regional resolutions and in discussions and negotiations on forest-related issues;

—examining the possibilities to take active measures to prevent and combat trade of illegally harvested wood;

—encouraging consideration of climate change effects in forestry;

(i) On genetically modified organisms (GMOs):

—developing the provisions and methods for risk assessment, identification, labeling and traceability of GMOs in order to enable effective monitoring and controls of health and environmental effects;

—aiming for swift ratification and implementation of the Cartagena Protocol on Biosafety and supporting the build up of regulatory frameworks in third countries where needed through technical and financial assistance.

Article 7

Objectives and priority areas for action on environment and health and quality of life

1. The aims set out in Article 2 should be pursued by the following objectives, taking into account relevant World Health Organization (WHO) standards, guidelines and programmes:

—achieving better understanding of the threats to environment and human health in order to take action to prevent and reduce these threats;

—contributing to a better quality of life through an integrated approach concentrating on urban areas;

—aiming to achieve within one generation (2020) that chemicals are only produced and used in ways that do not lead to a significant negative impact on health and the environment, recognizing that the present gaps of knowledge on the properties, use, disposal and exposure of chemicals need to be overcome;

—chemicals that are dangerous should be substituted by safer chemicals or safer alternative technologies not entailing the use of chemicals, with the aim of reducing risks to man and the environment;

—reducing the impacts of pesticides on human health and the environment and more generally to achieve a more sustainable use of pesticides as well as a significant overall reduction in risks and of the use of pesticides consistent with the necessary crop protection. Pesticides in use which are persistent or bio-accumulative or toxic or have other properties of concern should be substituted by less dangerous ones where possible;

—achieving quality levels of ground and surface water that do not give rise to significant impacts on and risks to human health and the environment, and to ensure that the rates of extraction from water resources are sustainable over the long term;

—achieving levels of air quality that do not give rise to significant negative impacts on and risks to human health and the environment;

—substantially reducing the number of people regularly affected by long-term average levels of noise, in particular from traffic which, according to scientific studies, cause detrimental effects on human health and preparing the next step in the work with the noise directive.

2. These objectives shall be pursued by means of the following priority actions:

(a) reinforcement of Community research programmes and scientific expertise, and encouragement to the international coordination of national research programmes, to support achievement of objectives on health and environment, and in particular the:

—identification and recommendations on the priority areas for research and action including among others, the potential health impacts of electromagnetic pollution sources and including particular attention to the development and validation of alternative methods to animal testing in particular in the field of chemical safety;

—definition and development of indicators of health and environment;

—re-examination, development and updating of current health standards and limit values, including where appropriate, the effects on potentially vulnerable groups, for example, children or the elderly, and the synergies and the reciprocal impact of various pollutants;

—review of trends and the provision of an early warning mechanism for new or emerging problems;

(b) On Chemicals:

—placing the responsibility on manufacturers, importers and downstream users for generating knowledge about all chemicals (duty of care) and assessing risks of their use, including in products, as well as recovery and disposal;

—developing a coherent system based on a tiered approach, excluding chemical substances used in very low quantities, for the testing, risk assessment and risk management of new and existing substances with testing procedures that minimize the need for animal testing and develop alternative testing methods;

—ensuring that the chemical substances of concern are subject to accelerated risk management procedures and that substances of very high concern, including carcinogenic, mutagenic or toxic for reproduction substances, and those which have POPs (persistent organic pollutants) characteristics, are used only in justified and must be subject to authorization before their use;

—ensuring that the results of the risk assessment of chemicals are taken fully into account in all areas of Community legislation where chemicals are regulated and to avoid duplication of work;

—providing criteria for including among the substances of very high concern those that are persistent and bio-accumulating and toxic and substances that are very persistent and very bio-accumulative and envisaging the addition of known endocrine disrupters when agreed test methods and criteria are established;

—ensuring that the main measures that are necessary in view of the identified objectives are developed speedily so that they can come into force before the mid-term review;

—ensuring public access to the non-confidential information in the Community Register on Chemicals (REACH Register);

(c) On pesticides:

—full implementation and review of the effectiveness of the applicable legal framework[10] in order to ensure a high level of protection, when amended. This revision might include, where appropriate, comparative assessment and the development of Community authorization procedures for placing on the market;

—a thematic strategy on the sustainable use of pesticides that addresses:

(i) Minimizing the hazards and risks to health and environment from the use of pesticides;

(ii) Improved controls on the use and distribution of pesticides;

(iii) Reducing the levels of harmful active substances including through substitution the most dangerous with safer, including non-chemical, alternatives;

(iv) Encouragement of the use of low input or pesticide free cultivation among others through raising users' awareness, promoting the use of codes of good practices, and promoting consideration of the possible application of financial instruments;

(v) A transparent system for reporting and monitoring progress made in fulfilling the objectives of the strategy including the development of suitable indicators;

(d) On chemicals and pesticides:

—aiming at swift ratification of the Rotterdam Convention on the Prior Informed Consent Procedure for Certain Hazardous Chemicals and Pesticides in International Trade and of the Stockholm Convention on Persistent Organic Pollutants (POPs);

—amending Council Regulation (EEC) No 2455/92 of 23 July 1992 concerning the export and import of certain dangerous chemicals[11] with the aim of bringing it into line with the Rotterdam Convention, improving its procedural mechanisms and improving information to developing countries;

—support the improvement of the management of chemicals and pesticides in developing and candidate countries, including the elimination of stocks of obsolete pesticides, *inter alia*, by supporting projects aimed at such elimination;

—contributing to international efforts on the elaboration of a strategic approach on international chemicals management;

(e) On the sustainable use and high quality of water:

—ensuring a high level of protection of surface and groundwater, preventing pollution and promoting sustainable water use;

—working towards ensuring full implementation of the Water Framework Directive,[12] aiming at a good ecological, chemical and quantitative water status and a coherent and sustainable water management;

—developing measures aimed at cessation of discharges, emissions and losses of Priority Hazardous Substances, in line with the provisions of the Water Framework Directive;

—ensuring a high level of protection of bathing water, including revising the Bathing Water Directive;[13]

—ensuring the integration of the concepts and approaches of the Water Framework Directive and of other water protection directives in other Community policies;

(f) On air quality, development and implementation of measure in Article 5 in the transport, industry and energy sectors should be compatible with and contribute to improvement on quality of air. Further measures envisaged are:

—improving the monitoring and assessment of air quality, including the deposition of pollutants, and the provision of information to the public, including the development and use of indicators;

—a thematic strategy to strengthen a coherent and integrated policy on air pollution to cover priorities for further actions, the review and updating where appropriate of air quality standards and national emission ceilings with a view to reach the long term objective of no-excedence of critical loads and levels and the development of better systems for gathering information, modeling and forecasting;

—adopting appropriate measures concerning ground-level ozone and particulates;

—considering indoor air quality and the impacts on health, with recommendations for future measures where appropriate;

—playing a leading role in the negotiations and the implementation of the Montreal Protocol on ozone depleting substances;

—playing a leading role in the negotiations on and strengthening the links and interactions with international processes contributing to clean air in Europe;

—further development of specific Community instruments for reducing emissions from relevant source categories;

(g) On noise:

—supplementing and further improving measures, including appropriate type-approval procedures, on noise emissions from services and products, in particular motor vehicles including measures to reduce noise from the interaction between tire and road surface that do not compromise road safety, from railway vehicles, aircraft and stationary machinery;

—developing and implementing instruments to mitigate traffic noise where appropriate, for example, by means of transport demand reduction, shifts to less noisy modes of transport, the promotion of technical measures and of sustainable transport planning;

(h) On urban environment:

—a thematic strategy promoting an integrated horizontal approach across Community policies and improving the quality of urban environment, taking into account progress made in implementing the existing cooperation framework[14] reviewing it where necessary, and addressing:

—the promotion of Local Agenda 21;

—the reduction of the link between economic growth and passenger transport demand;

—the need for an increased share in public transport, rail, inland waterways, walking and cycling modes;

—the need to tackle rising volumes of traffic and to bring about a significant decoupling of transport growth and GDP growth;

—the need to promote the use of low emission vehicles in public transports;

—the consideration of urban environment indicators.

Article 8

Objectives and priority areas for action on the sustainable use and management of natural resources and wastes

1. The aims set out in Article 2 should be pursued by the following objectives:

—aiming at ensuring that the consumption of resources and their associated impacts do not exceed the carrying capacity of the environment and breaking the linkages between economic growth and resource use. In this context the indicative target to achieve a percentage of 22% of the electricity production from renewable energies by 2010 in the Community is recalled with a view to increasing drastically resource and energy efficiency;

—achieving a significant overall reduction in the volumes of waste generated through waste prevention initiatives, better resource efficiency and a shift towards more sustainable production and consumption patterns;

—a significant reduction in the quantity of waste going to disposal and the volumes of hazardous waste produced while avoiding an increase of emissions to air, water and soil;

—encouraging re-use and for wastes that are still generated: the level of their hazardousness should be reduced and they should present as little risk as possible; preference should be given to recovery and especially to recycling; the quantity of waste for disposal should be minimized and should be safely disposed of; waste intended for disposal should be treated as closely as possible to the place of its generation, to the extent that this does not lead to a decrease in the efficiency in waste treatment operations.

2. These objectives shall be pursued taking into consideration the Integrated product Policy approach and the Community's strategy for waste management[15] by means of the following priority actions:

(i) developing a thematic strategy on the sustainable use and management of resources, including, *inter alia*:

(a) an estimate of materials and waste streams in the Community, including important ad exports for example, by using the instrument of material flow analysis;

(b) a review of the efficiency of policy measures and the impact of subsidies relating to natural resources and waste;

(c) establishment of goals and targets for resource efficiency and the diminished use of resources, decoupling the link between economic growth and negative environmental impacts;

(d) promotion of extraction and production methods and techniques to encourage eco-efficiency and the sustainable use of raw materials, energy, water and other resources;

(e) development and implementation of a broad range of instruments including research, technology transfer, market-based and economic instruments, programmes of best practice and indicators of resource efficiency;

(ii) Developing and implementing measure on waste prevention and management by, *inter alia*:

(a) developing a set of quantitative and qualitative reduction targets covering all relevant waste, to be achieved at Community level by 2010. The Commission is invited to prepare a proposal for such targets by 2002;

(b) encourage ecologically sound and sustainable product design;

(c) raising awareness of the public's potential contribution on waste reduction;

(d) The formulation of operational measures to encourage waste prevention, e.g., stimulating re-use and recovers, the phasing out of certain substances and materials through product-related measures;

(e) Developing further indicators in the field of waste management;

(iii) Developing a thematic strategy on waste recycling, including, *inter alia*:

(a) measures aimed at ensuring source separation, the collection and recycling of priority waste streams;

(b) further development of producer responsibility;

(c) development and transfer of environmentally sound waste recycling and treatment technology;

(iv) Developing or revising the legislation on wastes, including, *inter alia*, construction and demolition waste, sewage sludge,[16] biodegradable wastes, packing,[17] batteries[18] and waste shipments, clarification of the distinction between waste and non-waste and development of adequate criteria for the further elaboration of Annex IIA and IIB of the framework directive on wastes.

Article 9

Objective and priority areas for action on international issues

1. The aim set out in Article 2 on international issues and the international dimensions of the four environmental priority areas of this Programme involve the following objectives:

—the pursuit of ambitious environmental policies at the international level paying particular attention to the carrying capacity of the global environment;

—the further promotion of sustainable consumption and production patterns at the international level;

—making progress to ensure that trade and environment policies and measures are mutually supportive.

2. These objectives shall be pursued by means of the following priority actions:

(a) integrating environment protection requirements into all the Community's external policies, including trade and development cooperation, in order to achieve sustainable development by *inter alia* the elaboration of guidelines;

(b) establishing a coherent set of environment and development targets to be promoted for adoption as part of 'a new global deal or pact' at the World Summit on Sustainable Development in 2002;

(c) work towards strengthening international environmental governance by the gradual reinforcement of the multilateral cooperation and the institutional framework including resources;

(d) aiming for swift ratification, effective compliance and enforcement of international conventions and agreements relating to the environment where the Community is a Party;

(e) promoting sustainable environmental practices in foreign investment and export credits;

(f) intensify efforts at the international level to arrive at consensus on methods for the evaluation of risks to health and the environment, as well as approaches of risk management including the precautionary principle;

(g) achieving mutual supportiveness between trade and the need for environmental protection, by taking due account of the environmental dimension in Sustainability Impact Assessments of multilateral trade agreements to be carried out at an early stage of their negotiation and by acting accordingly;

(h) further promoting a world trade system that fully recognizes Multilateral or Regional Environmental Agreements and the precautionary principle, enhancing opportunities for trade in sustainable and environmentally friendly products and services;

(i) promoting cross-border environmental cooperation with neighboring countries and regions;

(j) promoting a better policy coherence by linking the work done within the framework of the different conventions, including the assessment of interlinkages between biodiversity and climate change, and the integration of biodiversity considerations into the implementation of the United Nations Framework Convention on Climate Change and the Kyoto Protocol.

Article 10

Environment policy making

The objectives set out in Article 2 on environment policy making based on participation and best available scientific knowledge and the strategic approaches set out in Article 3 shall be pursued by means of the following priority actions:

(a) development of improved mechanisms and of general rules and principles of good governance within which stakeholders are widely and extensively consulted at all stages so as to facilitate the most effective choices for the best results for the environment and sustainable development in regard to the measures to be proposed;

(b) strengthening participation in the dialogue process by environmental NGOs through appropriate support, including Community finance;

(c) improvement of the process of policy making through:

—*ex-ante* evaluation of the possible impacts, in particular of the environmental impacts, of new policies including the alternative of no action and of the proposals for legislation and publication of the results;

—*ex-post* evaluation of the effectiveness of existing measures in meeting their environmental objectives;

(d) ensuring that environment and notably the priority areas identified in this Programme are a major priority for Community research programmes. Regular reviews of environmental research needs and priorities should be undertaken within the context of the Community Framework Programme of research and technological development. Ensuring better coordination of research related to the environment conducted in Member States *inter alia* to improve the application of results;

Development of bridges between environmental and other actors in the fields of information, training, research, education and policies;

(e) ensuring regular information, to be provided starting from 203, that can help to provide the basis for:

—policy decisions on the environment and sustainable development;

—the follow-up and review of sector integration strategies as well as of the Sustainable Development Strategy;

—information to the wider public.

The production of this information will be supported by regular reports from the European Environment Agency and other relevant bodies. The information shall consist notably of:

—headline environmental indicators;

—indicators on the state and trends of the environment;

—integration indicators;

(f) reviewing and regularly monitoring information and reporting systems with a view to a more coherent and effective system to ensure streamlined reporting of high quality, comparable and relevant environmental data and information. The Commission is invited, as soon as possible, to provide a proposal as appropriate to this end. Monitoring, data collection and reporting requirements should be addressed efficiently in future environmental legislation;

(g) reinforcing the development and the use of earth monitoring (e.g., satellite technology) applications and tools in support of policy-making and implementation.

Article 11

Monitoring and evaluation of results

1. In the fourth year of operation of the Programme the Commission shall evaluate the progress made in its implementation together with associated environmental trends and prospects. This should be done on the basis of a comprehensive set of indicators. The Commission shall submit this mid-term report together with any proposal for amendment that it may consider appropriate to the European Parliament and the Council.

2. The Commission shall submit to the European Parliament and the Council a final assessment of the Programme and the state and prospects for the environment in the course of the final year of the Programme.

Article 12

This Decision shall be published in the Official Journal of the European Communities.

Done at Brussels, 22 July 2002.

For the European Parliament	*For the Council*
The President	*The President*
P. COX	P.S. MOLLER

NOTES

CHAPTER 1. INTRODUCTION

1. European Environment Agency, *Europe's Environment: The Third Assessment* (Copenhagen: EEA, 2003), 55.

2. European Environment Agency, *Europe's Environment: The Third Assessment* 56.

3. *Declaration of the United Nations Conference on the Human Environment,* Article 1, http://www.unep.org/Documents/Default.asp?Document ID=97&ArticleID=1503 Accessed 24 May 2003.

4. "Europe," Britannica Student Encyclopedia 2003 Encyclopedia Britannica Online. Accessed 4 March 2003. Available from http://search.eb.com/ebi/ article? eu=296100.

5. "Europe," Britannica Student Encyclopedia 2003 Online. Accessed 4 March 2003. Available from http://search.eb.com/ebi/article? eu=296100.

6. "Europe," Britannica Student Encyclopedia Online. Accessed 28 February 2003. Available from http://search.eb.com/ebi/article?eu=296100.

7. "Europe." Encyclopedia Britannica Online. Accessed 28 February 2003, http:// search.eb.com/eb/article?eu=33845.

8. "Europe." Encyclopedia Britannica Online. Accessed 28 February 2003. Available from http://search.eb.com/eb/article?eu=33845, 9.

9. UNESCO Institute for Statistics, "Literacy Rates." Accessed 21 April 2003. Available from http://portal.unesco.org/uis/ev.php?URL_ID=4926 &URL_ DO=DO_TOPIC&URL_SECTION=201.

10. "Europe," Encyclopedia Britannica Online. Accessed 28 February 2003. Available from http://search.eb.com/eb/article?eu=33845, 13.

11. "Europe," Encyclopedia Britannica. Accessed 28 February 2003. Available from http://search.eb.com/eb/article?eu=33845, 13.

12. "Europe," Encyclopedia Britannica Online. Accessed 28 February 2003. Available from http://search.eb.com/eb/article?eu=33845.

13. "Europe," Encyclopedia Britannica Online. Accessed 28 February 2003. Available from http://search.eb.com/eb/article?eu=33845.

14. "Europe," Encyclopedia Britannica Online. Accessed 28 February 2003. Available from http://search.eb.com/eb/article?3u=33845.

15. European Environment Agency, *Europe's Environment: The Second Assessment* (Oxford: Elsevier Science, 1998).

16. European Environment Agency, *Europe's Environment: The Second Assessment.*

17. European Environment Agency, *Europe's Environment: The Second Assessment.*

18. European Environment Agency, *Europe's Environment: The Second Assessment.*

19. European Environment Agency, *Europe's Environment: The Second Assessment.*

20. European Environment Agency, *Europe's Environment: The Second Assessment*, 109.

21. United Nations, "Report of the United Nations Conference on Environment and Development," Rio de Janeiro, 3–14 June 1992, Annex I Principle 4. Accessed 5 May 2003. Available from http:www.un.org/documents/ga/conf15126-1annex1.htm.

22. European Sustainable Cities and Towns Campaign, "Structure." Accessed 5 May 2003. Available from http://www.sustainable-cities.org/sub1main.html.

23. European Environment Agency, *Europe's Environment: The Second Assessment,* 269.

24. Legally the states were separate entities and are formally known as the European Communities.

25. As cited in Elinor Ostrom, *Governing the Commons: The Evolution of Institutions for Collective Action* (New York: Cambridge University Press, 1990), 69.

26. Publications such as Garrett Hardin's "Tragedy of the Commons" and Rachel Carson's *Silent Spring* played an important role in sensitizing people to environmental concerns.

27. United Nations Convention on the Law of the Sea, Convention on the Conservation of Antarctic Marine Living Resources, Convention on Long-range Transboundary Air Pollution, Convention on International Trade in Endangered Species, Convention on Wetlands on International Importance, Convention on Climate Change, Convention on Biological Diversity, Convention for the Protection of the Ozone Layer, and others.

28. James G. March and Johan P. Olsen, "The New Institutionalism: Organizational Factors in Political Life," *American Political Science Review* 78 (1984): 734–49.

29. Peter M. Haas, "Regime Patterns for Environmental Management," in Helge Hveem (Ed.), *Complex Cooperation: Institutions and Processes in International Resource Management* (Oslo, Norway: Scandinavian University Press, 1994), 39–59.

30. Robert O. Keohane, "Against Hierarchy: An Institutional Approach to International Environmental Protection," in Helge Hveem (Ed.), *Complex Cooperation: Institutions and Processes in International Resource Management* (Oslo, Norway: Scandinavian University Press, 1994), 13–34.

31. Robert L. Bish, "Environmental Resource Management: Public or Private?" in J. A. Baden and D. Noonan (Eds.), *Managing the Commons* (Indianapolis: Indiana University Press, 1998), 65–75.

32. Elinor Ostrom, *Governing the Commons: The Evolution of Institutions for Collective Action* (New York: Cambridge University Press, 1990).

33. As cited in E. Ostrom.

34. Robert O. Keohane, "Against Hierarchy."

35. E. Ostrom, *Governing the Commons*.

36. Mancur Olson, "Group Size and Group Behavior," in J. A. Baden and D. S. Noonan (Eds.), *Managing the Commons*, 1998), 39–50.

37. John A. Baden, "A New Primer for the Management of Common Pool Resources and Public Goods," in J. A. Baden and D. S. Noonan (Eds.), *Managing the Commons,* 51.

38. Oran R. Young (Ed.), *Global Governance: Drawing Insights from the Environmental Experience* (Cambridge, MA: MIT Press, 1997), 4.

39. Oran R. Young, *International Cooperation: Building Regimes for Natural Resources and the Environment* (Ithaca, NY: Cornell University Press, 1989).

40. John Vogler, *The Global Commons: A Regime Analysis* (New York: Wiley & Sons, 1995), 19.

41. Kenneth Hanf, "Implementing International Environmental Policies," in Andrew Blowers and Pieter Glasbergen (Eds.), *Environmental Policy in an International Context: Prospects* (New York: Wiley & Sons, 1996), 197–222.

42. Martin List, "Sovereign States and International Regimes," in Andrew Blowers and Pieter Glasbergen (Eds.), *Environmental Policy in an International Context: Prospects,* (New York: Wiley & Sons, 1996), 7–24.

43. Lamont C. Hempel, *Environmental Governance: The Global Challenge* (Washington, D.C.: Island Press, 1996).

44. World Commission on Environment and Development, *Our Common Future* (New York: Oxford University Press, 1987).

45. Elizabeth C. Economy and Miranda A. Schreurs, "Domestic and International Linkages in Environmental Politics" in Miranda A. Schreurs and Elizabeth C. Economy, (Eds.), *The Internationalization of Environmental Protection* (Cambridge, MA: Cambridge University Press, 1997). Pp. 1–18.

46. Elizabeth C. Economy, "Chinese Policy-Making and Global Climate Change: Two-Front Diplomacy and the International Community" in Miranda A. Schreurs and Elizabeth C. Economy. Pp. 19–41.

47. Joanne M. Kauffman, "Domestic and International Linkages in Global Environmental Politics: A Case Study of the Monteal Protocol," in Miranda A. Schreurs and Elizabeth C. Economy. Pp. 74–96.

48. Phyllis Mofson, "Zimbabwe and CITES: Illustrating the Reciprocal Relationship between the State and the International Regime," in Miranda A. Schreurs and Elizabeth C. Economy (eds.) pp. 162–87.

49. Gareth Porter and Janet Welsh Brown, *Global Environmental Politics* (Boulder, CO: Eastview Press, 1996).

50. Oran Young, *Global Governance: Drawing Insights from the Environmental Experience.*

51. Oran Young, *Global Governance.*

52. Lawrence E. Susskind, *Environmental Diplomacy: Negotiating More Effective Global Agreements* (NewYork, Oxford University Press, 1994); Elinor Ostrom, *Governing the Commons;* John Vogler. *The Global Commons.*

53. Oran Young, *Global Governance.*

54. Gareth Porter and Janet Welsh Brown, *Global Environmental Politics* (Boulder, CO: Eastviews, 1996).

55. John Vogler, *The Global Commons.*

56. John Vogler.

57. John Vogler.

58. Dwight R. Lee, "Environmental versus Political Pollution," in John A. Baden and Douglas S. Noonan (Eds.), *Managing the Commons* (Indianapolis: Indiana University Press, 1998).

59. Gareth Porter and Janet Welsh Brown, *Global Environmental Politics.*

60. Gareth Porter and Janet Welsh Brown.

61. D. G. Victor, K. Raustiala, and E. B. Skolnikoff (Eds.), *The Implementation and Effectiveness of International Environmental Commitments: Theory and Practice* (Cambridge, MA: MIT Press, 1998).

62. D. G. Victor, K. Raustiala, and E. B. Skolnikoff, 691.

63. Lawrence E. Susskind, *Environmental Diplomacy: Negotiating More Effective Global Agreements* (New York: Oxford University Press, 1994).

64. Lawrence E. Susskind.

65. Gareth Porter and Janet Welsh Brown, *Global Environmental Politics.*

66. Arild Underdal, "Measuring and Explaining Regime Effectiveness," in Helge Hveem (Ed.), *Complex Cooperation: Institutions and Processes in International Resource Management* (Oslo, Norway: Scandinavian University Press, 1994), 92–122.

67. Harold K. Jacobson and Edith Brown Weiss, "Strengthening Compliance with International Environmental Accords," in Paul F. Diehl (Ed.), *The Politics of Global Governance: International Organizations in an Interdependent World* (Boulder, CO: Lynne Rienner, 1997), 305–34.

68. D. G. Victor, K. Raustiala, and E. B. Skolnikoff (Eds.), *The Implementation and Effectiveness of International Environmental Commitments: Theory and Practice* (Cambridge, MA: MIT Press, 1998).

69. Peter Haas, "Regime Patterns for Environmental Management."

70. The Brundtland Report, UN World Commission on Environment and Development, 1987.

71. Lawrence E. Susskind, *Environmental Diplomacy.*

72. Lawrence E. Susskind.

73. Jacobson and Weiss, "Strengthening Compliance."

74. Michael Faure and Jürgen Lefevere, "Compliance with International Environmental Agreements," in Norman J. Vig and Regina S. Axelrod (Eds.), *The Global Environment: Institutions, Law, and Policy* (Washington, DC: CQ Press, 1999), 138–56.

75. Lawrence E. Susskind, *Environmental Diplomacy.*

CHAPTER 2. THE ORGANIZATIONAL STRUCTURE OF THE EUROPEAN UNION

1. Legally the states were separate entities and are formally called the "European Communities."

2. Bulgaria, Cyprus, the Czech Republic, Estonia, Hungary, Latvia, Lithuania, Malta, Poland, Romania, Slovakia, and Slovenia are currently negotiating to join the EU. Turkey is also a potential member.

3. *Treaty of Nice*, Official Journal C80/1. Accessed 1 April 2002. Available from http://www.europa.eu.int/eur-lex/en/treaties/dat/nice_treaty_en.pdf.

4. Any amendments to a treaty must be ratified by the member states according to their individual constitutional requirements.

5. Commission of the European Communities, *Communication from the Commission to the Council and the European Parliament, Biodiversity Action Plan for the Conservation of Natural Resources* (27 March, 2001), 21. Accessed 29 October 2001. Available from http://europa.eu.int/eur-lex/en/com/pdf/act0162en 01/com2001_0162en01_2.pdf.

6. Ian Ward, *A Critical Introduction to European Law* (London: Butterworths, 1996).

7. Euroconfidentiel, *The Rome, Maastricht and Amsterdam Treaties: Comparative Texts,* (Belgium: Euroconfidentiel, 1999), 41.

8. Article 228 discusses the making of agreements between the Community and one or more states or international organizations.

10. Clive Church and David Phinnemore, *European Union and European Community: A Handbook and Commentary on the Post-Maastricht Treaties* (New York: Harvester Wheatsheaf, 1994).

11. Euroconfidentiel, *The Rome, Maastricht and Amsterdam Treaties,* 41.

12. *Amsterdam Treaty: The Implications of Local Government.* Accessed 29 October 2001. Available from http://mhs.trinity-cm.ac.uk/europe/bp_al.htm.

13. European Commission, "The Amsterdam Treaty: A Comprehensive Guide." Accessed 15 October 2001. Available from http://europa.eu.int/scadplus/leg/en/lvb/a15000.htm.

14. *Consolidated Version of the Treaty on European Union,* 12. Accessed 27 March 20021. Available from http://www.europa.eu.int/eur-lex/en/treaties/ dat/eu_cons_treaty_en.pdf.

16. Treaty of Rome.

17. Treaty of Rome, Article 157.

18. European Commission, "The Role of the Commission." Accessed 28 March 2002. Available from http://www.europa.en.int/comm/role_en.htm#1.

19. European Commission, *Environment DG: Information Brochure* (Luxembourg: Office for Official Publications of the European Communities, 2002).

20. European Commission, 4.

21. Council of the European Union, "Qualified Majority." Accessed 31 March 2002. Available from http://ue.eu.int/en/Info/index.htm.

22. European Commission, *The Institutions and Bodies of the European Union* (Luxembourg: Office for Official Publications of the European Communities, 2001).

23. European Commission, *The Institutions and Bodies of the European Union*.

24. The third covers provisions relating to fiscal matters.

25. European Commission, *The Institutions and Bodies of the European Union*.

26. Klaus-Dieter Borchardt, *The ABC of Community Law* (Brussels: European Communities, 2000). Accessed 25 March 2002. Available from http://www.europa.eu.int/eur-lex/en/about/abc_en.pdf.

27. Euroconfidentiel, *The Rome, Maastricht and Amsterdam Treaties*. "The European Parliament." Accessed 31 March 2002. Available from http://www.europarl.eu.int/presentation/default_en.htm .

28. "Rules of Procedure of the European Parliament, Chapter XXIII Petitions, Rule 174 Right of Petition" (14th ed., June 3, 1999). Accessed 31 March 2002. Available from http://www2.europarl.eu.int/omk/OM-Europarl?LEVEL=2& SAME_LEVEL=1&PROG=RULES-EP&LINK=1&L=EN&PUBREF=-//EP//TEXT+RULES-EP+-+RULE-174+DOC+SGML+V0//EN.

29. *Rules of Procedure of the European Parliament,* Chapter XXIII Petitions, Rule 175 Examination of petitions" (14th ed, June 3, 1999). Accessed 31 March 2002. Available from http://www2.europarl.eu.int/omk/OM-Europarl?LEVEL=0&SAME_LEVEL=1&PROG=RULES-EP&LINK=1&L=EN&PUBREF=-//EP//TEXT+RULES-EP+-+RULE-175+DOC+SGML+V0//EN.

30. As provided for in EC Treaty Article 190 (138)1)).

31. "The European Parliament: Electoral Procedures." Accessed 25 March 2002. Available form http://www.europarl.eu.int/factsheets/1_3_4_en.htm.

32. "The European Parliament: Electoral Procedures."

33. Europarl, "Members of the European Parliament." Accessed 23 October 2002. Available from http://www.eruoparl.eu.int/presentation/default_en.htm.

34. "The European Parliament: Electoral Procedures."

35. Europarl, "Members of the European Parliament."

36. European Commission, *The Institutions and Bodies of the European Union*.

37. European Parliament, "Members of the European Parliament, 5th Term, 1999–2004." Accessed 21 October 2002. Available from http://wwwdb.europarl.eu.int/ep5/owa/p_meps2.repartition?ipid=1103391&ilg=EN&iorig=home&imsg=.

38. "A Guide to the European Parliament," <http://www.europarl.org.uk/guide/GEPmain.htm> 26 March 2002.

39. There are also six advocates-general, who are appointed in the same manner as the judges.

40. Clive Church and David Phinnemore, *European Union and European Community*.

41. Court of Justice, "Recent Case Law," Accessed 22 October 2001. Available from http://curia.eu.int/jurisp/html/text.htm.

42. European Council, "The European Council's Major Political Role in Developing the European Union." Accessed 23 October 2001. Available from http://ue.eu.int/en/info/eurocouncil/sommet.htm.

43. Treaty of Rome, Articles 193 and 197.

44. Treaty of Rome, Article 198.

45. "Treaty of Nice, Amending the Treaty on European Union, the Treaties Establishing the European Communities and Certain Related Acts," *Official Journal of the European Communities*, C 80/1 10.3.2001. Acccessed 28 March 2002. Available from http://europa.eu.int/eur-lex/en/treaties/dat/nice_treaty_en.pdf .

46. Euroconfidentiel, *The Rome, Maastricht, and Amsterdam Treaties*, 177.

47. Euroconfidentiel, *The Rome, Maastricht, and Amsterdam Treaties*, 177.

48. Committee of the Regions "Introduction." Accessed 22 October 2001. Available from http://www.cor.eu.int/presentation/prxrol00_en.htm.

49. Committee of the Regions, "Role." Accessed 27 March 2002. Available from http://www.cor.eu.int.

50. Euroconfidentiel, *The Rome, Maastricht, and Amsterdam Treaties*, 180.

51. Committee of the Regions, "Summary." Accessed 31 March 2002. Available from http://www.cor.eu.int/corz101.htm.

52. European Union, "Institutions of the European Union." Accessed 15 October 2001. Available from http://europa.eu.int/inst-en.htm.

53. The European ombudsman, "Decisions concerning Environment." Accessed 24 March 2002. Available from http://www.euro-ombudsman.eu.int/decision/en/env.htm.

54. Council Directive of 27 June 1985 on the assessment of the effects of certain public and private projects on the environment (85/337/EEC), Official Journal L 175, 5.7.1985, 40, amended by Council Directive 97/11/Ec of 3 March 1997, Official Journal L 73, 14.3.1997, 5. Accessed 27 March 2002. Available from http://europa.eu.int/eur-lex/en/consleg/main/1985/en_1985L0337_index.html.

55. "Proposal for a Directive of the European Parliament and of the Council Establishing a Scheme for Greenhouse Gas Emission Allowance Trading within the Community and Amending Council Directive 96/61/EC." Accessed 27 March 2002. Available from http://europa.eu.int/eur-lex/en/com/dat/2001/en_501PC0581.html.

56. Opinion of the Committee of the Regions on the "Proposal from the Commission for a Directive Providing for Public Participation in Respect of the Drawing Up of Certain Plans and Programmes Relating the Environment and Amending Council Directives 85/337/EEC and 96/61/EC," Official Journal C 357/58 14.12.2001. Accessed 26 March 2002. Available from http://europa.eu.int/eur-lex/pri/en/oj/dat/2001/c_357200112/4en00580060.pdf.

57. Document 390R1210: Council Regulation (EEC) No. 1210/09 of 7 May 1990 on the establishment of the European Environment Agency and the European Environment Information and Observation Network, Official Journal NO. L 120, 11/05/1990, 0001–06. Accessed 22 October 2001. Available from http://eruopa.eu.int/eur-lex/en/lif/dat/1990/en_390R1210.htm.

58. European Environment Agency, *Annual Report 1998*, 9.

59. Council Regulation (EEC) No. 1210/90 of 7 May 1990 on the establishment of the European Environment Agency and the European Environment Information and Observation Network, 10/22/01. Accessed 15 October 2001. Available from http://europa.eu.int/eur-lex/en/lif/dat/1990/en_390R1210.html.

60. European Environment Agency, "List of Members of the Management Board." Accessed 25 October 2001. Available from http://org.eea.eu.int/organisation/manboard.shtml.

61. European Environment Agency, *Annual Report 1998*, 10.

62. Joachim Scherer Baker and Frankfurt McKenzie, "Environmental Regulation of the European Community," in *Regulating the European Environment,* Thomas Handler (Ed.) (New York: Wiley & Sons, 1997).

63. European Environmental Protection Agency, "Towards Sustainable Development—The Evolution of EU Environmental Policy." Accessed 20 October 2001. Available from http://europa.eu.int/pol/environment/info_en.htm.

64. European Environment Agency, *Annual Report 1998,* 10.

65. European Environment Agency, *The DPSIR Framework.* Accessed 22 October 2001. Available from http://reports.eea.eu.int:80/SPE19961113/en/page 005.html.

66. European Environment Agency, *Annual Report 1998,* 12.

67. Alexandre Kiss and Dinah Shelton, *Manual of European Environmental Law.*

68. European Commission, "Eurobarometer 58.0—The Attitudes of Europeans towards the Environment." Accessed 22 May 2003. Available from http://europa.eu.int/comm/environment/barometer/index.htm.

69. EU Commission, Environment Directorate, "The Aarhus Convention." Accessed 22 May 2003. Available from http://europa.eu.int/comm/environment/aarhus/index.html.

70. European Commission, "Communication from the Commission: Towards a Reinforced Culture of Consultation and Dialogue—General Principles and Minimum Standards for Consultation of Interested Parties by the Commission, Brussels, 11.12.2002, COM(2002) 704 final, 6. Accessed 22 May 2003. Available from http://europa.eu.int/eur-lex/en/com/cnc/2002/com2002_0704en02.pdf .

71. EU Commission, "Your Voice in Europe." Accessed 22 May 2003. Available from http://europa.eu.int/yourvoice/index_en. htm.

72. European Commission, "What Is Interactive Policy Making?" Accessed 22 May 2003. Available from http://ipmmarkt.homestead.com/.

73. Philip Raworth, *The Legislative Process in the European Community* (Boston: Kluwer Law and Taxation Publishers, 1993).

74. European Commission, "Communication from the Commission."

CHAPTER 3. THE EUROPEAN ENVIRONMENT

1. For example, migratory birds have come to depend on human made structures, such as chimneys. Land management that did not take into consideration those human intrusions in nature would be disruptive to the ecosystems that have developed.

2. Ian Ward, *A Critical Introduction to European* Law (London: Butterworths, 1996), 22.

3. European Union, "Towards Sustainable Development—The Evolution of EU Environmental Policy." Accessed 20 March 2002. Available from http://europa.eu.int/pol/env/info_en.htm.

4. Commission of the European Communities, *Communication from the Commission to the Council and the European Parliament, Biodiversity Action Plan*

for the Conservation of Natural Resources (27 March 2001, 23). Accessed 29 October 2001. Available from http://europa.eu.int/eur-lex/en/com/pdf/act0162en01/com2001_0162en01_2.pdf.

5. For a history of the EIS process in the United States see Matthew Lindstrom and Zachary A. Smith, *The National Environmental Policy Act* (Albany: State University of New York Press, 2001).

6. G. D. Burholt and A. Martin, *The Regulatory Framework for Storage and Disposal of Radioactive Waste in the Member States of the European Community* (London: Graham & Trotman, 1988).

7. Euroconfidentiel, *The Rome, Maastricht, and Amsterdam Treaties: Comparative Texts* (Belgium, Euroconfidentiel, 1999), 41.

8. Clive H. Church and David Phinnemore, *European Union and European Community: A Handbook and Commentary on the Post-Maastricht Treaties* (New York: Harvester Wheatsheaf, 1994).

9. Euroconfidentiel, *The Rome, Maastricht, and Amsterdam Treaties*, 41.

10. Committee of the Regions, "Opinion of the Committee of the Regions on the Revision of the Treaty on European Union." (Brussels, 20 April 1995). Accessed 25 March 2002. Available from http://europa.en.int/en/agenda/igc-home/eu-doc/regions/crf_en.html.

11. European Policy Center, "Beyond the Delimitation of Competences: Implementing Subsidiarity: *The Europe We Need* Working Paper." (25 September 2001). Accessed 24 March 2002. Available from http://europa.eu.int/futurum/documents/other/oth250901_en.pdf .

12. Ute Collier, *Energy and Environment in the European Union* (Brookfield, VT: Ashgate, 1994).

13. François Lévêque, "Introduction," in François Lévêque (Ed.), *Environmental Policy in Europe: Industry, Competition and the Policy Process* (Brookfield, MA: Elgar, 1996), 8.

14. François Lévêque, 15.

15. European Commission, "Preparation of the New Environmental Action Programme." Accessed 23 October 2002. Available from http://europa.eu.int/comm/environment/newprg/preparation.htm.

16. J. D. Liefferink, P. C. Lowe, and A. P. J. Mol, "The Environment and the European Community: The Analysis of Political Integration," in J. D. Liefferink, P. C. Lowe, and A. P. J. Mol (Eds.), *European Integration and Environmental Policy* (London: Belhaven, 1993), 4.

17. "Resolution of the Council of the European Communities and of the Representatives of the Governments of the Member States, Meeting within the Council of 19 October 1987 on the Continuation and Implementation of a European Community Policy and Action Programme on the Environment (1987–1992)," Official Journal C 328, 07/12/1987, 7. Accessed 30 March 2002. Available from http://europa.eu.int/eur-lex/en/lif/dat/1987/en_487Y1207_01.htm.

18. Andrew Blowers and Pieter Glasbergen (Eds.), *Environmental Policy in an International Context: Prospects* (New York: Wiley & Sons, 1996).

19. "EEC Treaty, Article 100A." Accessed 29 March 2002. Available from http://europa.eu.int/comm/employment_social/h&s/intro/art100a_en.htm.

20. "Resolution of the Council of the European Communities and of the Representatives of the Governments of the Member States, Meeting within the Council of 19 October 1987 on the Continuation and Implementation of a European Community Policy and Action Programme on the Environment (1987–1992)," Official Journal C 328, 07/12/1987, 0001–0044. Accessed 30 March 2002. Available from http://europa.eu.int/eur-lex/en/lif/dat/1987/en_487Y1207_01.html.

21. "Resolution of the Council of the European Communities and of the Representatives of the Governments of the Member States, Meeting within the Council of 19 October 1987 on the Continuation and Implementation of a European Community Policy and Action Programme on the Environment (1987–1992)," Official Journal C 328, 07/12/1987, 7. Accessed 30 March 2002. Available from http://europa.eu.int/eur-lex/en/lif/dat/1987/en_487Y1207_01.html.

22. "Towards Sustainability," the European Community Program of Policy and Action in Relation to the Environment and Sustainable Development (better known as the Fifth EC Environmental Action Program). Accessed 29 March 2002. Available from http://europa.eu.int/comm/environment/actionpr.htm.

23. European Commission, "Towards Sustainability," the Fifth Environmental Action Program." Accessed 13 May 2003. Available from http://europa.eu.int/comm/environment/actionpr.htm=.

24. Commission of the European Communities, *Communication from the Commission to the Council and the European Parliament: Biodiversity Action Plan for the Conservation of Natural Resources*, Brussels: COM(2001) 162 final, Vol. II (7 March 2001). Accessed 25 October 2001. Available from http://europa.eu.int/eur-lex/en/com/pdf/act0162en01/com2001-01623n01-2.pdf.

25. Brian Rothery, *What Maastricht Means for Business* (Brookfield, VT: Gower, 1993).

26. European Commission, "The Global Assessment." Accessed 18 May 2003. Available from http://europa.eu.int/comm/environment/newgprg/global.htm.

27. European Communities, "Environment 2010: Our Future, Our Choice: The Sixth Environment Action Programme of the European Community 2001–2010." Accessed 30 October 2001. Available from http://www.europa.eu.int/comm/environment/newprg/index.htm.

28. Decision No. 1600/2002/EC of the European parliament and of the council of 22 July 2002 laying down the Sixth Community Environment Action Program, Article 1, (Official Journal of the European Communities), 10.9. 2002, L242/3.

29. Commission of the European Communities, *Communication from the Commission to the Council, the European Parliament, the Economic and Social Committee and the Committee of the Regions on the Sixth Environment Action Programme of the European Community: Environment 2010: Our Future, Our Choice*, COM (2001) 31 final. Brussels, 24 January 2001, 5.

30. Decision No. 1600/2002/EC, Article 6.

31. European Environment Agency, "EU Greenhouse Gas Emissions Rise for Second Year Running," Accessed 23 may 2003. Available from http://www.eea.eu.int/.

32. European Commission, Environment Directorate, "The CAFÉ Programme." Accessed 26 May 2003. Available from http://europa.eu.int/comm/environment/air/cafe/index.htm.

CHAPTER 4. SUCCESSES AND CHALLENGES IN EUROPEAN UNION ENVIRONMENTAL POLICY

1. BirdLife International, *Action Plan for Fea's Petrel.* (London: BirdLife International on behalf of the European Commission, April 1996).

2. European Commission, *Council Direction 79/409/EEC on the Conservation of Wild Birds.* Accessed 2 December 2002. Available from http://europa.eu.int/comm/environment/nature/bird-dir.htm.

3. Commission of the European Communities, Report form the Commission on the Application of Directive 79/409/EEC on the Conservation of Wild Birds, Update for 1993–99 (Brussels: European Commission, 20.02.2000), COM (2000) 180 final.

4. "In Focus: The NATURA 2000 Network," *NATURA 2000 Newsletter* 1 (May 1998). Accessed 11 December 2002. Available from http://europa.eu.int/comm/environment/news/natura//nat1_en.htm.

5. Additional information on the international program can be found at http://www.redlist.org/info/programme.html. Accessed 3 March 2003.

6. BirdLife International, *Action Plan for Fea's Petrel* (London: BirdLife International on behalf of the European Commission, April 1996.)

7. European Commission, *LIFE-Nature: A Brief History of Nature Conservation Financing.* Accessed 23 December 2002. Available from http://europa.eu.int/comm/environment/life/life/nature_history.htm.

8. European Commission, *Welcome to the LIFE Homepage.* Accessed 23 December 2002. Available from http://europa.eu.int/comm/environment/life/home.htm.

9. European Commission, *LIFE-III: The Financial Instrument for the Environment.* Accessed 23 December 2002. Available from http://europa.eu.int/comm/environment/life/life/index.htm.

10. BirdLife International, *Birds to Watch 2.* Accessed 30 November 2002. Available from http://www.wcmc.org.uk/species/data/red_note/12930.htm.

11. Patrick Ten Brink, Claire Monkhouse, and Saskia Richartz, *Promoting the Socio-Economic Benefits of Natura 2000.* (London: Institute for European Environmental Policy, November 2002).

12. *Geography: Spain.* Accessed 7 December 2002. Available from http://geography.about.com/library/cia/lbespain.htm.

13. "European Union-Structural Funds." Accessed 3 March 2003. Available from http://midlands.ie/content/6eu/6_11.htm.

14. European Commission, "Introducing the ESF." Accessed 3 March 2003. Available from http://europa.eu.intcomm/employment_social/esf2000/introduction-en.htm.

15. Calculated of exchange rates in this chapter were done online at the December 2002 exchange rate of 1 euro = US $1.03; Accessed 26 December 2002. Available from www.expedia.com/pub/agent.dll.

16. Patrick Ten Brink, Claire Monkhouse, and Saskia Richartz, *Promoting the Socio-Economic Benefits of Natura 2000,* 19.

17. European Commission, *Support for Rural Development.* Accessed 24 December 2002. Available from http://europa.eu.int/scadplus/leg/en/lvb/l60062.htm 24.

18. European Commission, *Support for Rural Development.* Accessed 24 December 2002. Available from http://europa.eu.int/scadplus/leg/en/lvb/l60062.htm.

19. *Denmark,* Microsoft Encarta Online Encyclopedia 2002. Accessed 12 December 2002. Available from http://encarta.msn.com.

20. "Abolishing Pesticides Could Prove *Expensive*," *The Copenhagen Post* (19 November 1998). Accessed 12 December 2002. Available from http://cphpost.periskop.dk/default.asp?id=8863.

21. European Environment Agency, *Europe's Environment: The Second Assessment* (Oxford: Elsevier Science, 1998), 190.

22. Philippa Blincoe, "Ambitious Eco Targets: Honorable but Not Effective?" *The Copenhagen Post (* 4 September 1999). Accessed 12 December 2002. Available from http://cphpost.periskop.dk/default.asp?id=8863.

23. European Environment Agency, *Europe's Environment,* 183.

24. However, there are currently 100,106 substances that can be used without testing, and the burden of proof of harm is on public authorities.

25. Ralf Nordbeck and Michael Faust, *UFZ Discussion Papers: European Chemicals Regulation and Its Effect on Innovation*, UFZ Center for Environmental Research, Department of Economics, Sociology and Law, Leipzig, Germany. Accessed 12 December 2002. Available from http://eeb.org/publication/Innovations.orrkungen_eng.pdf.

26. European Commission, *Community Water Policy.* Accessed 23 December 2002. Available from http://europa.eu.int/scadplus/leg/en/lvb/l280029.htm.

27. European Commission, *Framework Directive in the Field of Water Policy.* Accessed 23 December 2002. Available from http://europa.eu.int/scadplus/leg/en/lvb/l28002b.htm.

28. European Commission, *Framework Directive in the Field of Water Policy,* Accessed 23 December 2002. Available from http://europa.eu.int/scadplus/leg/en/lvb/l28002b.htm.

29. John Hontelez, Secretary General, European Environmental Bureau, *Letter to the Ministers for the Environment of the EU Member States*, 29 November 2002.

30. European Commission, *Pollution Caused by Nitrates from Agricultural Sources.* Accessed 24 December 2002. Available from http://europa.eu.int/scadplus/leg/en/lvb/l28013.htm.

31. European Commission, *Pollution Caused by Nitrates from Agricultural Sources.* Accessed 24 December 2002. Available from http://europa.eu.int/scadplus/leg/en/lvb/l28013.htm.

32. Median income calculated from data provided at *Incomes, consumption, and prices: Family income, form of tenure and use of private cars 1996.* www2.dst.dk/internet/klb/dod99/STYR9.HTM 26 December 2002 and an exchange rate calculated online at the December 2002 rate of 1DKK = US $.14, Accessed 26 December 2002. Available from www.expedia.com/pub/agent.dll.

33. *Geography: Finland*, Accessed 27 December 2002. Available from http://geography.about.com/library/ cia/lbefinland.htm.

34. European Commission, "Transport: Motorization: Number of Passenger Cars per 1000 Inhabitants." Accessed 28 December 2002. Available from

http://europa.eu.int/comm/energy_transport/etif/transport_means_road/motorization.htm.

35. Finland Ministry of the Environment, "Environmental Legislation." Accessed 12 December 2002. Available from http://www.vyh.fi/eng/environ/legis/index.htm.

36. Finnish Ministry of the Environment, "Finland's Indicators for Sustainable Development: Taxes per Carbon Dioxide Content of Fuels." Accessed 12 December 2002. Available from http://www.vyh.fi/eng/environ/sustdev/indicat/poltto.htm.

37. European Commission, "Environment: CO2 Emissions from Fossil Fuels by Sector (EU 15)." Accessed 28 December 2002. Available from http://europa.eu.int/comm/energy_transport/etif/environment/emissions_sector_table.html.

38. European Commission, "Motor Vehicles with Trailers: Petrol and Diesel Engines." Accessed 23 December 2002. Available from http://europa.eu.int/scadplus/leg/en/lvb/l21047.htm.

39. European Commission, *Air Quality*. Accessed 24 December 2002. Available from http://europa.eu.int/comm/environment/air/index.htm.

40. European Commission, *CO_2 Emissions from New Passenger Cars: Monitoring*, http://europa.eu.int/scadplus/leg/en/lvb/l28055.htm 23 December 2002.

41. European Commission, "Objective of the Work on Fiscal Framework Measures to Reduce CO_2 Emissions from Passenger Cars." Accessed 24 December 2002. Available from http://europa.eu.int/comm/environment/ CO_2/ CO_2_expgrp.htm.

42. European Commission, "Environment and Energy." Accessed 6 June 2003. Available from http://europa.eu.int/comm/environment/integration/energy_en.htm.

43. European Commission, "Energy: The Integration of Environmental Considerations into the Energy Sector." Accessed 8 June 2003. Available from http://europa.eu.int/comm/environment/environment-act5/chapt1-2.htm.

44. European Commission, "Environmental Technology Action Plan." Accessed 8 June 2003. Available from http://europa.eu.int/comm/environment/etap/index.htm.

45. "Wind Farm Development—Guidelines for Planning Authorities." Accessed 27 May 2003. Available from http://www.environ.ie/search/searchindex.html.

46. European Commission, *LIFE-Environment in Action: 56 New Success Stories for Europe's Environment* (Luxembourg: Office for Official Publications of the European Communities, 2001).

47. "Wind Farm Development—Guidelines for Planning Authorities." Accessed 27 May 2003. Available from http://www.environ.ie/search/searchindex.html.

48. European Commission, "Multiannual Programme for Action in the Field of Energy." Accessed 7 June 2003. Available from http://europa.eu.int/scadplus/leg/en/lvb/l27046.htm.

49. European Commission, "Multiannual Programme for Action in the Field of Energy." Accessed 7 June 2003. Available from http://europa.eu.int/scadplus/leg/en/lvb/l27046.htm.

50. European Commission, "Renewable Energies—Altener." Accessed 27 May 2003. Available from http://europa.eu.int/scadplus/leg/en/lvb/l27016b.htm.

51. European Commission, "Renewable Energies—Altener." Accessed 27 May 2003. Available from http://europa.eu.int/scadplus/leg/en/lvb/l27016b.htm.

52. Council of the European Union, "A Strategy for Integration Environmental Aspects and Sustainable Development into Energy Policy." Accessed 6 June 2003. Available from htttp://europa.eu.int/comm./energy_transport/library/council-resol-environment.pdf.

53. European Union, "Renewable Energy: White Paper Laying Down a Community Strategy and Action Plan." Accessed 27 May 2003. Available from Sustainable Energy Ireland, "RE Energy Supply in Europe." Accessed 27 May 2003. Available from http://www.environ.ie/search/searchindex.htmlAvailablromttp://www.environ.ie/search/searchindex.html.

54. Seas at Risk, *The Future of European Fisheries: Recommendations for Reform of the Common Fisheries Policy*. Accessed 20 November 2002. Available from http://www.seas-at-risk.org/SAR_recommendations_for_the_CFP_review.doc.

55. Seas at Risk, *The Future of European Fisheries: Recommendations for Reform of the Common Fisheries Policy.* Accessed 20 November 2002. Available from http://www.seas-at-risk.org/SARrecommendations_for_THE_cfp_review.doc, 2.

56. European Commission, *Green Paper on the Future of the Common Fisheries Policy*, COM(2001) 135. Accessed 20 March 2001 Available from http://europa.eu.int/eur-lex/en/com/gpr/2001/com2001_0135en01.pdf 3 December 2002, 7.

57. European Commission, *Project No. 96/030: Study of Exploited Fish Stocks on the Flemish Cap.* Accessed 3 December 2002. Available from http://europa.eu.int/comm/fisheries/doc_et_publ/liste_publi/studies/biological/1309R03B96030.pdf.

58. Alex Kirby, "UK Cod Fishing 'Could Be Halted.'" Accessed 3 December 2002. Available from http://news.bbc.co.uk/1/hi/sci/techy/1005565.stm.

59. European Commission, *Helping the Recovery of Cod and Northern Hake.* Accessed 11 December 2002. Available from hpttp:europa.eu.int/comm./fisheries/topics/topic_en.htm.

60. European Commission, *Green Paper on the Future of the Common Fisheries Policy*, COM(2001) 135, (20 March 2001). Accessed 3 December 2002. Available from http://europa.eu.int/eur-lex/en/com/gpr/2001/com2001_0135en01.pdf.

61. European Commission, *Management of Fisheries Resources*. Accessed 11 December 2002. Available from http://europa.eu.int.comm./fisheries/faq/manage_en.htm.

62. Niki Sporrong, "CFP Compromise on the Cards," *El Anzuelo: European Newsletter on Fisheries and the Environment* (London: Institute for European Environmental Policy, 2002), Vol. 10, 2.

63. BBC News, "Finnie Fights 'Draconian' Cod Cuts." Accessed 3 December 2002. Available from http://news.bbc.co.uk/1/hi/scotland/2512003.stm.

64. *Transcript in English of Internet Chat with Commissioner Franz Fischler on the Reform of the Common Fisheries Policy.* Accessed 3 December 2002. Available from http://europa.eu.int/comm/chat/fischler2/fischler2_en.pdf.

65. European Commission, *Structural Measures*. Accessed 1 December 2002. Available from http://europa.eu.int/comm/fisheries/faq/structures_en.htm.

66. Commission of the European Communities, *Eighteenth Annual Report on Monitoring the Application of Community Law,* Brussels, 16.7.2001 COM (2001)309 final, Vol. 1. Accessed 15 November 2002. Available at http://europa.eu.int/eur-lex/en/com/rpt/2001/act309en01/com2001_0309en01-01.pdf.

67. Commission of the European Communities, *Eighteenth Annual Report on Monitoring the Application of Community Law,* Brussels, 16.7.2001 COM (2001)309 final, Vol. 1. Accessed 15 November 2002. Available from http://europa.eu.int/eur-lex/en/com/rpt/2001/act309en01/com2001_0309en01-01.pdf, 47.

68. Commission of the European Communities, *Eighteenth Annual Report on Monitoring the Application of Community Law,* Brussels, 16.7.2001 COM (2001)309 final, Vol. 1. Accessed 15 November 2002. Available from http://europa.eu.int/eur-lex/en/com/rpt/2001/act309en01/com2001_0309en01-01.pdf, 49.

69. Commission of the European Communities, *Eighteenth Annual Report on Monitoring the Application of Community Law,* Brussels, 16.7.2001 COM (2001)309 final, Vol. 1. Accessed 15 November 2002. Availavle from http://europa.eu.int/eur-lex/en/com/rpt/2001/act309en01/com2001_0309en01-01.pdf, 47.

70. "Opinions of the Advocate-General delivered 26 February 1992, Commission of the European Communities v Hellenic Republic," *European Court Reports 1992, Page I–02509*. Accessed 4 February 2003. Available from http://europa.eu.int/smartapi/cgi/sga_doc?smartapi!celexapi!prod!CELEXnumdoc&lg=EN&numdo.

71. Based on exchange rate of US $1 to 0.852023 euros as of 7 June 2003. Calculated at Expedia.com Exchange Rate Calculator. Accessed 7 June 2003. Available from http://www.expedia.com/pub/agent.dll.

72. Greece Starts to Pay Huge Toxic Waste Dump Fine," *Environment News Service*. Accessed 4 February 2003. Available from http://ens-news.com/ens/jan2001/2001-01-10-02.asp.

73. Judgment of the court of 4 July 2000, *European Commission v Hellenic Republic*. Accessed 4 February 2003. Available from http://eruopa.eu.int/smartapi/cgi/sga_doc?smartapi!celexapi!prod!CELEXnumdoc&lg=EN&numdo.

74. Commission of the European Communities, *Eighteenth Annual Report on Monitoring the Application of Community Law,* Brussels, 16.7.2001 COM (2001)309 final, Vol. 1. Accessed 4 February 2003. Available from http://europa.eu.int/eur-lex/en/com/rpt/2001/act309en01/com2001_0309en01-01.pdf.

75. Lectric Law Library. Accessed 7 June 2003. Available from http://www.lectlaw.com/def2/q101.htm.

76. Commission of the European Communities, *Eighteenth Annual Report on Monitoring the Application of Community Law,* Brussels, 16.7.2001 COM (2001)309 final, Vol. 1. Accessed 15 November 2002. Available from http://europa.eu.int/eur-lex/en/com/rpt/2001/act309en01/com2001_0309en01-01.pdf, 76.

77. "The Role and Scope of IMPEL." Accessed 26 May 2003. Available from http://europa.eu.int/comm/environment/impel/role_scope.pdf.

78. "The Role and Scope of IMPEL." Accessed 26 May 2003. Available from http://europa.eu.int/comm/environment/impel/role_scope.pdf.

79. European Commission, "About IMPEL," http://europa.eu.int/comm/environment/impel/about.htm 26 May 2003.

80. IMPEL Network, *IMPEL: Best Practice in Compliance Monitoring,* 18–21 June 2001. Accessed 24 May 2003. Available from http://europa.eu.int/comm/environment/impel/compliance.pdf .

81. European Commission, "About IMPEL." Accessed 26 May 2003. Available from http://europa.eu.int/comm/environment/impel/about.htm.

82. "IRI Ireland." Accessed 26 May 2003. Available from http://europa.eu.int/comm/environment/impel/iri_ireland.htm.

83. European Commission, "About IMPEL," Accessed 26 May 2003. Available from http://europa.eu.int/comm/environment/impel/about.htm.

84. "IRI Ireland." Accessed 26 May 2003. Available from http://europa.eu.int/comm/environment/impel/iri_ireland.htm.

85. European Environment Agency, *Europe's Environment: The Third Assessment* (Copenhagen: EEA, 2003), 5. Accessed 25 May 2003. Available from http://reports.eea.eu.int/environmental_assessment_report_2003_10/en.

CHAPTER 5. A COMPARATIVE EVALUATION OF THE EUROPEAN UNION AND THE UNITED STATES

1. *Oxford Atlas of the World* (New York: Oxford University Press, 1997); U.S. Census Bureau, *Statistical Abstract of the United States, 2000.* Accessed 15 July 2002. Available from http://www.census.gov/prod/2001pubs/statab/sec06.pdf; metric conversion from square miles to square kilometers performed at "Convertme.com," http://www.convert-me.com/en/convert/area 4 March 2003.

2. United States Census Bureau, "Time Series of National Population Estimates." Accessed 18 July 2002. Available from http://eire.census.gov/popest/data/national/populartables/table01.php; European Commission, "Eurostat General Statistics." Accessed 15 July 2002. Available from http://europa.eu.int/comm/eurostat/Public/datashop/print-catalogue/EN?catalogue=Eurostat&theme=1-General%20Statistics.

3. U.S. Environmental Protection Agency, "About EPA." Accessed 12 November 2002. Available from http://www/epa.gov/epahome/abouteps.htm.

4. U.S. Environmental Protection Agency, "Strategic Plan" (Washington, DC: Office of the Chief Financial Officer, 2002). Accessed 13 November 2002. Available from http://www.epa.gov/ocfo/plan/2000strategicplan.pdf.

5. U.S. Environmental Protection Agency, "Strategic Plan" (Washington, DC: Office of the Chief Financial Officer, 2002), 10–11. Accessed 13 November 2002. Available from http://www.epa.gov/ocfo/plan/2000strategicplan.pdf.

6. U.S. Environmental Protection Agency, "Global Warming—Actions." Accessed 13 November 2002. Available from http://yosemite.epa.gov/OAR/globalwarming.nsf/content/ActionsInternationalOutreachInteresting.html.

7. United Nations Environment Program and the Climate Change Secretariat, *Understanding Climate Change: A Beginner's Guide to the UN Framework Convention and Its Kyoto Protocol.* Accessed 14 November 2002. Available from http://unfccc.int/resource/beginner_02-en.pdf.

8. Jason F. Shogren and Michael A. Toman, "Climate Change Policy," in *Public Policies for Environmental Protection*, Paul R. Portney and Robert N. Stavins (Eds.) (Washington, DC: Resources for the Future Press, 2000), 128.

9. Intergovernmental Panel on Climate Change, *Climate Change 1995: The Science of Climate Change*, Contribution of Working Group I to the Second Assessment Report of the Intergovernmental Panel on Climate Change (New York: Cambridge University Press, 1996); National Research Council, *Reconciling Observations of Global Temperature Change* (Washington, DC: National Academy Press, 2000).

10. Climate Change Secretariat, *A Guide to the Climate Change Convention and Its Kyoto Protocol*. Accessed 14 November 2002. Available from http://unfccc.int/resource/guideconv/cp-p.pdf.

11. International Institute for Environment and Development, *Development: The Challenge of Climate Change: EC Aid and Sustainable Development Briefing Paper, No. 6,* 1995, 2. Accessed 14 November 2002. Available from http://europa.eu.int/comm/development/sector/environment/env_+theme/climate_change/documents/06.htm.

12. U.S. Environmental Protection Agency, "Global Warming—Emissions." Accessed 12 November 2002. Available from http://yosemite.epa.gov/oar/global-warming.nsf/content/emissionsindividual.htm.

13. U.S. Environmental Protection Agency, "Global Warming—Emissions." Accessed 12 November 2002. Available from http://yosemite.epa.gov/oar/global-warming.nsf/content/emissionsindividual.htm.

14. *Global Climate Change Policy Book (2002)*. Accessed 11 November 2002. Available from http://yosemite.epa.gov/oar/globalwarming.nsf/UniqueKeyLookup/SHSU5BNMAJ/$File/bush_gccp_021402.pdf; White House, "Fact Sheet: President Bush Announces Clear Skies and Global Climate Change Initiatives," 3. Accessed 11 November 2002. Available from http://www.whitehouse.gov/news/releases/2002/02/20020214.html.

15. *Global Climate Change Policy Book (2002)*. Accessed 11 November 2002. Available from http://yosemite.epa.gov/oar/globalwarming.nsf/UniqueKeyLookup/SHSU5BNMAJ/$File/bush_gccp_021402.pdf; White House, "Fact Sheet: President Bush Announces Clear Skies and Global Climate Change Initiatives," 3. Accessed 11 November 2002. Available from http://www.whitehouse.gov/news/releases/2002/02/20020214.html.

16. Environmental Protection Agency, "Current US Actions to Address Climate Change." Accessed 12 November 2002. Available from http://yosemite.epa.gov/oar/globalwarming.nsf/UniqueKeyLookup/SHSU5BNM7H/$File/bush_ccpol_061101.pdf.

17. Environmental Protection Agency, "Current US Actions to Address Climate Change." Accessed 12 November 2002. Available from http://yosemite.epa.gov/oar/globalwarming.nsf/UniqueKeyLookup/SHSU5BNM7H/$File/bush_ccpol_061101.pdf.

18. White House, "Fact Sheet: President Bush Announces Clear Skies and Global Climate Change Initiatives." Accessed 11 November 2002. Available from http://www.whitehouse.gov/news/releases/2002/02/20020214.html.

19. Robert J. Guttman, "Europe Interview," *Europe,* 418 (July/August 2002), 26–27.

20. European Commission, "Strategy for Sustainable Development." Accessed 23 October 2002. Available from http://europa.eu.int/scadplus/leg/en/lvb/128117.htm.

21. European Commission, *Commission Staff Working Paper: Third Communication from the European Community under the UN Framework Convention on Climate Change, 30 November 2001* (Brussels: 12 December 2001, SEC(2001) 2053. Accessed 14 November 2002. Available from http://europa.eu.int/comm/environment/climat/official_sec_2001_2053_en.pdf.

22. European Commission, *Commission Staff Working Paper: Third Communication from the European Community under the UN Framework Convention on Climate Change, 30 November 2001* (Brussels: 12 December 2001, SEC(2001) 2053. Accessed 14 November 2002. Available from http://europa.eu.int/comm/environment/climat/official_sec_2001_2053_en.pdf.

23. March Consulting Group, *Opportunities to Minimize Emissions of Hydrofluorocarbons (HFCs) from the European Union, Final Report, 30 September 1998.* Accessed 14 November 2002. Available from http://europa.eu.int/comm.enterprise/chemicals/sustdev/studhfc.pdf.

24. "Current US Actions to Address Climate Change." Accessed 12 November 2002. Available from http://yosemite.epa.gov/oar/globalwarming.nsf/UniqueKeyLookup/SHSU5BNM7H/$File/bush_ccpol_061101.pdf.

25. U.S. Environmental Protection Agency, "CERCLA Overview." Accessed 10 November 2002. Available from http:www.epa.gov/superfund/action/law/cercla.htm.

26. Environmental Protection Agency, "Introduction to the HRS." Accessed 12 November 2002. Available from http://www.epa.gov/superfund/programs/npl_hrs.hrsint.htm.

27. Environmental Protection Agency, "How Sites Are Placed on the NPL." Accessed 10 November 2002. Available from http://www/epa.gov/superfund/programs/npl_hrs/nplon.htm.

28. Katherine N. Probst, Don Fullerton and Robert E. Litan, *Footing the Bill for Superfund Cleanups: Who Pays and How?* (Washington, DC: Brookings Institution, Resources for the Future, 1995).

29. U.S. Environmental Protection Agency, "SARA Overview." Accessed 10 November 2002. Available from http://www.epa.gov/superfund/action/law/sara.htm.

30. U.S. Environmental Protection Agency, "National Contingency Plan Overview." Accessed 11 November 2002. Available from http://www.epa.gov/oil-spill/ncpover.htm.

31. Commission of the European Communities, *Proposal for a Directive of the European Parliament and of the Council on Environmental Liability with Regard to the Prevention and Remedying of Environmental Damage*, (Brussels: 2002; COM/0021(COD). Accessed 11 November 2002. Available from http://europa.eu.int/eur-lex/en/ com/pdf/2002/en_502PC0017.pdf.

32. Commission of the European Communities, *Proposal for a Directive of the European Parliament and of the Council on Environmental Liability with Regard to the Prevention and Remedying of Environmental Damage* (Brussels: 2002; COM/0021(COD), 2. Accessed 11 November 2002. Available from http://europa.eu.int/eur-lex/en/com/pdf/2002/en_502PC0017.pdf.

33. *Handbook for Implementation of EU Environmental Legislation—Water.* Accessed 14 November 2002. Available from http://europa.eu.int/comm/environment/enlarg/handbook/water.pdf.

34. Commission of the European Communities, *Proposal for a Directive of the European Parliament and of the Council on Environmental Liability with Regard to the Prevention and Remedying of Environmental Damage* (Brussels: 2002; COM/0021(COD), 4. Accessed 11 November 2002. Available from http://europa.eu.int/eur-lex/en/com/pdf/2002/en_502PC0017.pdf 11 November 2002, p.4.

35. European Commission, *An Analysis of the Preventive Effect of Environmental Liability.* Accessed 14 November 2002. Available from http://www.europa.eu.int/comm/environment/liability/preventive.htm.

36. Commission of the European Communities, *Proposal for a Directive of the European Parliament and of the Council on Environmental Liability with Regard to the Prevention and Remedying of Environmental Damage* (Brussels: 2002; COM/0021(COD), 3. Accessed 11 November 2002. Available from http://europa.eu.int/eur-lex/en/com/pdf/2002/en_502PC0017.pdf.

37. Mark O'Donovan, "Environmental Liability: Striking a Balance," *Environment for Europeans*, 10 (April 2002).

38. "Donana National Park." Accessed 12 November 2002. Available from http://whc.unesco.org/sites/685.htm.

39. "Destinations: Donana National Park." Accessed 12 November 2002.l Available from http://gorp.com/gorp/europe/sp_park/donana.htm.

40. United Press International, "Toxic Waste Threatens Europe's Largest Nature Reserve." Accessed 12 November 2002. Available from http://forests.org/archive/europe/toxwaste.htm.

41. Earth Crash, "Documenting the Collapse of a Dying Planet." Accessed 12 November 2002. Available from http://www.eces.org/articles/static/945064800 24555.shtml.

42. Ralf Nordbeck and Michael Faust, *UFZ Discussion Papers: European Chemicals Regulation and Its Effect on Innovation*, UFZ Center for Environmental Research, Department of Economics, Sociology and Law, Leipzig, Germany. Accessed 12 December 2002. Available from http://eeb.org/publication/Innovations.orrkungen_eng.pdf.

43. European Commission, "Chemicals." Accessed 11 December 2002. Available from http://europa.eu.int/comm/environment/chemicals/index.htm.

44. Jim Glen, "The State of Garbage," *Biocycle* (April 1992), 46

45. Conservation Foundation, *State of the Environment: A View towards the Nineties* (Washington, DC: Conservation Foundation, 1987), 11.

46. 42 U.S.C.A. Section 6943 (a) (2).

47. European Commission, "Waste Disposal." Accessed 11 January 2003. Available from http://europa.eu.int/scadplus/leg/en/lvb/121197.htm.

48. European Commission, "Landfill of Waste." Accessed 11 January 2003. Available from http://europa.eu.int/scadplus/leg/en/lvb/121208.htm.

49. European Commission, "Management of End-of-life Vehicles." Accessed 11 January 2003. Available from http://europa.eu.int/scadplus/leg/en/lvb/121225.htm.

50. European Environment Agency, *Europe's Environment: The Second Assessment*, (Luxembourg: Elsevier Science, 1998), 238.

51. European Commission, "Environment and Agriculture." Accessed 10 October 2002. Available from http://europa.eu.int/comm/environment/agriculture/index.htm.

52. European Commission, *Agriculture and Forestry.* Accessed 21 July 2002. Available from http://europa.eu.int/comm./environment/env_act5/chapt1-4.htm.

53. European Commission, *Communication from the Commission to the Council, the European Parliament, the Economic and Social Committee, and the Committee of the Regions on the Sixth Environment Action Programme of the European Community* (Brussels, 24 January 2001, COM (2001) 31 final. Accessed 10 October 2002. Available from http://europa.eu.int/eur-lex/en/com/pdf/2001/en_501 PC0031.pdf.

54. 33 U.S.C. Section 1362 (14).

55. European Commission, "Community Water Policy." Accessed 11 January 2003. Available from http://europa.eu.int/scadplus/leg/en/lvb/128002a.htm.

56. "Council Directive 98/83/EC on the Quality of Water Intended for Human Consumption" (OJ L330, 5.12.98). Accessed 12 August 2002. Available from http://europa.eu.int/comm/environment/enlarg/handbook/water.pdf.

57. European Commission, "Pricing and Long-term Management of Water." Accessed 11 January 2003. Available from http://europa.eu.int/scadplus/leg/en/lvb/128112.htm.

58. European Commission, "Pricing and Long-term Management of Water." Accessed 11 January 2003. Available from http://europa.eu.int/scadplus/leg/en/lvb/128112.htm.

59. D. Stanners and P. Bourdeau, (Eds.), *Europe's Environment: The Dobris Assessment* (Copenhagen: European Environment Agency, 1995).

60. United Nations, "Sustainable Development: Agenda 21." Accessed 9 January 2002. Available from http://www.un.org/esa/sustdev/agenda21.htm.

61. United Nations, "Sustainable Development: Agenda 21, Chapter 28: Local Authorities' Initiatives in Support of Agenda 21." Accessed 11 January 2002. Available from http://www.un.org/esa/sustdev/agenda21chapter28.htm.

62. European Environment Agency, *Europe's Environment: The Second Assessment*, (Luxembourg: Elsevier Science, 1998), 265.

63. European Environment Agency, *Europe's Environment*, 265.

64. John De Graff, David Wann, and Thomas H. Naylor, *Affluenza* (San Francisco: Berrett-Koehler, 2001).

CHAPTER 6. CONCLUSIONS

1. Douglas S. Noonan, "International Fisheries Management Institutions: Europe and the South Pacific," in J. A. Baden and D. S. Noonan (Eds.), *Managing the Commons* (Indianapolis: Indiana University Press, 1998), 154–65.

2. Oran R. Young (Ed.), *Global Governance: Drawing Insights from the Environmental Experience* (Indianapolis: Indiana University Press, 1998), 39–50.

3. Oran R. Young, *International Cooperation: Building Regimes for Natural Resources and the Environment* (Ithaca, NY: Cornell University Press, 1989).

APPENDIX 1. ARTICLES 249–256 OF THE TREATY ON EUROPEAN UNION

1. Euroconfidential, *The Rome, Maastricht and Amsterdam Treaties: Comparative Texts* (Brussels: Euroconfidential, 1999).

APPENDIX 2. ARTICLES 174–176 OF THE TREATY ON EUROPEAN UNION

1. Euroconfidential, *The Rome, Maastricht and Amsterdam Treaties: Comparative Texts* (Brussels: Euroconfidential, 1999).

2. Article 95 refers to consultations with the Economic and Social Committee related to the establishment and functioning of the internal market.

APPENDIX 3. THE SIXTH COMMUNITY ENVIRONMENT ACTION PROGRAM

1. "Decision No 1600/2002/EC of the European Parliament and of the Council of 22 July 2002 Laying Down the Sixth Community Environment Action Programme." (Official Journal of the European Communities, 10.9.2002, L242, 1–15).

2. Official Journal (OJ) C 154E, 29.5.2001, 218.

3. Official Journal C 221, 7.8.2001, 80.

4. Official Journal C 357, 14.12.2001, 44.

5. Opinion of the European Parliament of 31 May 2001 (OJ C 47 E, 21.2.2002, 113), Council Common Position of 27 September 2001 (OJ C 4, 7.1.2002, 52.) and Decision of the European Parliament of 17 January 2002 (not yet published in the Official Journal). Decision of the European Parliament of 30 May 2002 and Decision of the Council of 11 June 2002.

6. Regulation (EC) NO 761/2001 of the European Parliament and of the Council of 19 March 2001 allowing voluntary participation by organizations in a Community eco-management and audit scheme (EMAS) (O) L 114, 24.4.2001, 1.

7. Regulation (EC) No 1655/2001 of the European Parliament and of the Council of 17 July 2000 concerning the Financial Instrument for the Environment (LIFE) (O) L 192, 28.7.2000, 1.

8. Convention on Access to Information, Public Participation in Decision-Making and Access to Justice in Environmental Matters, Aarhus, 25 June 1998.

9. Official Journal L 309, 13.12.1993, 1.

10. Council Directive 91/414/EEC of 15 July 1991 concerning the placing of plant protection products on the market (OJ L 230, 19.8.1991, 1). Directive as last amended by Commission Directive 2001/49/EC (OJ L 176, 29.6.2001, 61).

11. Official Journal L 251, 29.8.1992, 13. Regulation as last amended by Commission Regulation (EC) No 2247/98 (OJ L 282, 20.10.1998, 12).

12. Directive 2000/60/EC of the European Parliament and of the Council of 23 October 2000 establishing a framework for Community action in the field of water policy (OJ L 327, 22.12.2000, 1).

13. Council Directive 76/160/EEC of 8 December 1975 concerning the quality of bathing water (OJ L 31, 5.2.1976, 1). Directive as last amended by the 1994 Act of Accession.

14. Decision No 1141/2001/EC of the European Parliament and of the Council of 27 June 2001 on a Community framework for cooperation to promote sustainable urban development (OJ L 191, 13.7.2001, 1).

15. Council Resolution of 24 February 1997 on a Community strategy for waste management (OJ C76, 11.3.1997,1).

16. Council Directive 86/278/EEC of 12 June 1986 on the protection of the environment, and in particular of the soil, when sewage sludge is used in agriculture (OJ L 181, 4.7.1986, 6.) Directive as last amended by the 1994 Act of Accession.

17. Directive 94/62/EC of the European Parliament and of the Council of 20 December 1994 on packaging and packaging waste (OJ L 365, 31.12.1994, 10). Directive as last amended by Commission Decision 1999/177/EC (OJ L 56, 4.3 1999, 47).

18. Commission Directive 93/86/EEC of 4 October 1993 adapting to technical progress Council Directive 91/157/EEC on batteries and accumulators containing certain dangerous substances (OJ L 264, 23.10.1993, 51).

INDEX

A Critical Introduction to European Law, 171, 174
ABC of Community Law, The, 172
Acidification, 8
ACNAT. *See* Actions by the Community for Nature
Action Plan for Fea's Petrel, 176
Action Programs, Environmental. See Environmental Action Programs
Administrative Procedures Act, United States, 94
Affluenza, 186
Agricultural Guidance and Guarantee Fund, European, 76
ALTENER, 81, 82, 84
Amsterdam Treaty. *See* Treaty of Amsterdam
Amsterdam Treaty: The Implications of Local Government, 171
Animals. *See* Bonn Convention on the Protection of Migratory Animals
Annual Report, 1998, European Environment Agency, 173–74
Article 130, Treaty on European Union, 30–31
Assent, 38, 39
Atomic Energy Community, European, 27
Axelrod, Regina S., 170

Baden, John A., 169–70, 187
Barcelona Convention for the Protection of the Marine Environment and the Coastal Region of the Mediterranean, 60
Bathing water. *See* Council Directive 76/160/EEC concerning the quality of bathing water
Benelux Union, 27
Bern Convention on the Conservation of European Wildlife and Natural Habitats, 60
Biodiversity Action Plan for the Conservation of Natural Resources, Communication from the Commission to the Council and the European Parliament, 171, 74–76
Biodiversity, 10, 171, 174–76
Birdlife International, 176
Birds to Watch 2, 177
Bish, Robert, 169
Blincoe, Philippa, 178
Blowers, Andres, 169, 175
Bonn Convention on the Protection of Migratory Animals, 60
Borchardt, Klaus-Dieter, 172
Bourdeau, P., 186
Brown Weiss, Edith, 170
Brundtland Report. *See Our Common Future*
Bureau of Land Management, United States, 96
Burholt, G. D., 175
Bush, George W., 98, 101, 118
Bush, George, 98

Carbon dioxide, 81–82
Carson, Rachel, 104, 168
Cartagena Biosafety Protocol, 60
CERCLA *See* United States Comprehensive Environmental Response, Compensation, and Liability Act
Chemicals, 9, 77–79
Chlorofluorocarbons, 99
Church Clive, 171–72, 175
CITES *See* Convention on International Trade in Endangered Species of Wild Fauna and Flora
Clean Air Act, United States, 96, 98, 115
Clean Water Act, United States, 95–96, 114–15
Clear Skies Initiative, 102
Climate Change Program, European, 103
Climate change, 3, 80
Clinton, William, 98
Coal and Steel Community, European, 27
Codecision, 38, 39
Codes of conduct, non-binding, 22
Collective management. *See* cooperative management
Collier, Ute, 175
Commission v Greece, C–45–01, 88
Committee of the Regions, 33, 36, 43–44, 46, 61, 123, 173
Common Agricultural Policy, 62, 113
Common Fisheries Policy, 85–87
Common Fisheries Policy, Green Paper on the Future of the, 85, 180
Common pool environmental resources, 14
Community for Nature, Actions by, 74
Community Framework for cooperation to promote sustainable urban development, 188
Complex Cooperation, 168
Complex Cooperation: Institutions and Processes in International Resource Management, 170
Comprehensive Environmental Response, Compensation, and Liability Act, United States, 104–7, 109–10
Conservation Foundation, 185
Conservation of Wild Birds, Council Directive 709/409/EEC, 177
Consolidated Version of the Treaty on European Union, 171
Consultation, 38, 39
Convention for the Protection of the Rhine, 60
Convention on Access to Information, Public Participation in Decision-Making and Access to Justice in Environmental Matters, United Nations, 54
Convention on Biodiversity, 60
Convention on Environment and Development, United Nations, 117
Convention on International Trade in Endangered Species of Wild Fauna and Flora, 18
Convention on the Human Environment, United Nations, 2, 58
Cooperation, 38, 39
Cooperative management, 15–16
Council of Ministers, 33, 34, 36, 54
Council of the European Union. *See* Council of Ministers
Court of Justice. *See* European Court of Justice

De Graff, John, 186
Decisions, 37, 38
Declaration of the United Nations Conference on the Human Environment, 167
Department of Agriculture, United States, 96
Directives, 37, 38
Directorate-general of the Environment. *See* Environment Directorate-General
Dobris Assessment, 117
Doñana National Park, 106, 109, 185

DPSIR Framework, 174
Drinking Water Directive, European Union, 115

Earth Crash, 185
Eastland Woolen Mill, 107
Eco-labels, 67
Economic and Social Committee, 33, 36, 42–43, 44, 123
Economy, Elizabeth C., 169
Eighteenth Annual Report on Monitoring the Application of Community Law, 181
EIONET. *See* European Information and Observation Network
El Anzuelo: European Newsletter on Fisheries and the Environment, 180
End-of-life Vehicles, Directive on, 112
Energy and Environment in the European Union, 175
Energy Supply, Green Paper on the Security of, 103
Environment 2010: Our Future, Our Choice, 68
Environment Directorate-General, 34, 37, 55
Environment for Europeans, 185
Environment, definition, 1
Environmental Action Programs, 62–69, 124
Environmental Diplomacy: Negotiating More Effective Global Agreements, 170
Environmental impact statement (Directive 97/11/EC), 59
Environmental Liability, Proposal for, 106
Environmental Policy in an International Context, 169, 175
Environmental Policy in Europe: Industry, Competition and the Policy Process, 175
Environmental Protection Agency Hazard Ranking System, United States, 105–6
Environmental Protection Agency, United States, 95–6, 98, 104, 183–84
Environmental resources, 14
Environmental responsibility, Finland, 80
Environmental Technology Action Plan, 81
EPA. *See* United States Environmental Protection Agency
Erika oil tanker, 106, 109
ETAP. *See* Environmental Technology Action Plan
EU Heads of State or Government, 33, 42, 172
Euratom. *See* European Atomic Energy Community
Euroconfidential, 173, 187
Europe environmental problems, 7–12
Europe's Environment: The Dobris Assessment, 186
Europe's Environment: The Second Assessment, 168, 178, 186
Europe's Environment: The Third Assessment, 167, 182
European Assembly. *See* European Parliament
European Commission, 33, 58, 114, 123, 172, 174, 176–79, 184–86
European Communities, 168, 171
European Council, 33
European Court of Justice, 33, 41, 45, 88, 124, 172
European Economic Community, 27–28
European Environment Agency, 49–52, 53, 71, 117, 125, 173, 175, 186
European Information and Observation Network, 51
European Integration and Environmental Policy, 175
European Ombudsman, 33, 45–46
European Union and European Community: A Handbook and Commentary on the Post-Maastricht Treaties, 171, 175
European Union Environmental Action Programs, 3, 103
European Union institutional environment, 3

European Union Network for the Implementation and Enforcement of Environmental Law, 90–91
European Union, members, 12

Faure, Michael, 170
Faust, Michael, 178, 185
Fea's Petrel, 73–5
Federal Insecticide, Fungicide, and Rodenticide Act, United States, 96
Financial Instrument for the Environment, 74
Fisheries, 90. *See* also Common Fisheries Policy
Footing the Bill for Superfund Cleanups: Who Pays and How? 184
Forest Glen Mobile Home Subdivision, 107
Forest Service, United States, 96
Framework Convention on Climate Change, United Nations, 60
Framework conventions, 22
Framework Directive in the Field of Water Policy, *178*
Fullerton, Don, 184
Future of European Fisheries: Recommendations for Reform of the Common Fisheries Policy, The, 180

Glasbergen, Pieter, 169, 175
Glen, Jim, 185
Global Assessment, 68
Global Change Research Act of 1990, United States, 103
Global Change Research Program, United States, 103
Global Climate Change Initiative, 102
Global Commons: A Regime Analysis, 169–70
Global Environment: Institutions, Law, and Policy, The, 170
Global Environmental Politics, 169–70
Global Governance: Drawing Insights from the Environmental Experience, 169–70
Governing the Commons, 168–70
Greenhouse effect, 99
Greenhouse gas emissions. *See* Council Directives 85/337/EEC and 96/61/EC on Public Participation in the Development of Plans Relating to the Environment, 47–48
Guttman, Robert J, 184

Haas, Peter, 168, 170
Habitats Directive 92/42/EEC, 49, 50, 74
Handbook for Implementation of EU Environmental Legislation—Water, 185
Handler, Thomas, 174
Hanf, Kenneth, 169
Hardin, Garrett, 168
Hazardous Waste, Community Directive on, 49
Helsinki Convention on the Protection of the Marine Environment of the Baltic Sea, 60
Hemple, Lamont C., 169
Hveem, Helge, 170

IMPEL. *See* European Union Network for the Implementation and Enforcement of Environmental Law
Implementation and Effectiveness of International Environmental Commitments: Theory and Practice, 170
Infrastructure for Spatial Information in Europe, 55
INSPIRE. *See* Infrastructure for Spatial Information in Europe
Institutionalism, 14–15
Institutions and Bodies of the European Union, The, 171–72
Intergovernmental Panel on Climate Change, 99–100, 183
International Cooperation: Building Regimes for Natural Resources and the Environment, 169, 187

International Institute for Environment and Development, 183
International organizations, 17
International regimes, 17
Internationalization of Environmental Protection, The, 169
IPCC. *See* Intergovernmental Panel on Climate Change

Jacobson, Harold K., 170
Joule-Thermie, 81

Karffman, Joanne M., 169
Keohane, Robert, 168, 169
Kiss, Alexandre, 174
Kyoto Protocol on Climate Change, 60, 101–2, 118

Lee, Dwight R., 170
Lefevere, Jurgen, 170
Legislative Process in the European Community, The, 174
Letter to the Ministers for the Environment of the EU Member States, 178
Lévêque, François, 175
Liefferink, J. D., 175
LIFE. *See* Financial Instrument for the Environment
LIFE-III: The Financial Instrument for the Environment, 177
LIFE-Nature: A Brief History of Nature Conservation Financing, 177
Lindstrom, Matthew, 175
List, Martin, 169
Litan, Robert E., 184
Local Agenda 21, 117
Lowe, P. C., 175

Maastricht Treaty. *See* Treaty on European Union
Managing the Commons, 169–70, 170, 187
Manual of European Environmental Law, 174
March Consulting Group, 184
March, James, 168
Martin, A., 175
McKenzie, Frankfurt, 174
Methane, 99
Mofson, Phyllis, 169
Mol, A. P. J., 175
Monkhouse, Claire, 177

National Environmental Policy Act, The, 175
National Oil and Hazardous substances Pollution Contingency Plan, United States, 106
National Priorities List, United States, 105
National Research Council, 99–100
Natura 2000 Newsletter, 177
NATURA 2000, 55, 74, 75
Natural hazards, 11–12
Natural resources, definition, 1
Naylor, Thomas H., 186
NEPA *See* United States National Environmental Policy Act
Nitrates, 77–78
Nitrous oxides, 99
Noonan, Douglas S., 179, 187
Nordbeck, Ralf, 178, 185

O'Donovan, Mark, 185
Official Journal of the European Communities, 173, 187–88
Oil Pollution Act, United States, 106
Olsen, Johan, 168
Olson, Mancur, 169
Opinions, 27
Ostrom, Elinor, 168–70
Our Common Future, 18, 23, 169

Packaging. *See* Directive 2000/60/EC of the European Parliament and of the Council establishing a framework for Community action in the field of water policy, 188
Paris Conference of the Heads of States and Governments, 58

Parliament, European, 33, 36, 39–41, 123, 172
Pesticides. *See* Chemicals
Phinnemore, David, 171–72, 175
Politics of Global Governance: International Organizations in an Interdependent World, The, 170, 187
Porter, Gareth, 169–70
Portney, Paul R., 183
Pricing and Sustainable Management of Water Resources, Communication from the Commission to the Council, European Parliament, and Economic and Social Committee regarding, 115–16
Privatization, 15
Probst, Katherine N., 184
Promoting the Socio-Economic Benefits of Natura 2000, 177
Proposal for a Directive of the European Parliament and of the Council on Environmental Liability with Regard to the Prevention and Remedying of Environmental Damage, 185
Protocols, 22
Public management, 16–17
Public Participation in the Development of Plans Relating to the Environment. *See Public Policies for Environmental Protection,* 183
Public. *See* Council Directives 85/337/EEC and 96/61/EC on Public Participation in the Development of Plans Relating to the Environment

Qualified majority, 36–37
Quality of Bathing Water, Council Directive on, 188
Quality of Water Intended for Human Consumption, Council Directive on, 115

Raustiala, K., 170
Raworth, Philip, 174
RCRA. *See* Resource Conservation and Recovery Act
Reagan, Ronald, 98
Recommendations, 37, 38
Regime effectiveness, 22–23
Regulating the European Environment, 174
Regulations, 37, 38
Regulatory Framework for Storage and Disposal of Radioactive Waste in the Member States of the European Community, 175
Renewable Energy Information Office, Ireland, 83
Resource Conservation and Recovery Act, United States, 96, 111–12, 114
Riaza River Gorges, 75
Richartz, Saskia, 177
Rio Earth Summit, 100
Rio World Summit. *See* Framework Convention on Climate Change
Rome, Maastricht, and Amsterdam Treaties, The, 173
Rothery, Brian, 176
Rules of Procedures of the European Parliament, 39, 172
Rural Development *See* Council Regulation 1257/1999 on Support for Rural Development

Safe Drinking Water Act, United States, 96, 114–15
Salzburg Initiative, 24
SARA. *See* Superfund Amendments and Reauthorization Act
Save and Save II, 81
Scherer Baker, Joachim, 174
Schreurs, Miranda A., 169
Seas at Risk, 180
Sewage, Council Directive 86/278/EEC on the protection of the Environment and in particular of the soil, when sewage sludge is used in agriculture, 188
Shelton, Dinah, 174
Shogren, Jason F., 183
Silent Spring, 104, 168
Single European Act, 28–29, 58

Sixth Environment Action Programme of the European Community: Environment 2010: Our Future, Our Choice, Communication from the Commission to the Council, the European Parliament, the Economic and Social Committee and the Committee of the Regions on the, 176
Sixth European Environmental Action Program, 103, 136–66
Skolnikoff, E. B., 170
Smigh, Zachary A., 175
Social Fund, European, 75–76
Soft law. *See* nonbinding codes of conduct
Solid Waste Disposal Act, United States, 112
Sporrong, Niki, 180
Stanners, D., 186
State of the Environment: A View towards the Nineties, 185
Stavins, Robert N., 183
Stock pollutants, 99
Stockholm Conference. *See* United Nations Conference on the Human Environment
Stockholm Convention on Persistent Organic Pollutants, 60
Stratospheric ozone level, 2, 8
Study of exploited Fish Stocks on the Flemish Cap, Project No. 96/030, 180
Subsidiarity, 30, 61, 63–64
Superfund Amendments and Reauthorization Act, United States, 105–6
Superfund. *See* Comprehensive Environmental Response, Compensation, and Liability Act; Superfund Amendments and Reauthorization Act
Superfund. *See* United States Comprehensive Environmental Response, Compensation, and Liability Act
Support for Rural Development, 178
Support for Rural Development, Council Regulation, 76
Supranational management. *See* International regimes
Susskind, Lawrence E., 170
Sustainable agriculture, 114
Sustainable Cities and Towns Campaign, European,168
Sustainable Development Strategy, 106
Sustainable development, 32
Synergie, 81

Technological hazards, 11–12
Ten Brink, Patrick, 177
The Internationalization of Environmental Protection, 169
The Rome, Maastricht and Amsterdam Treaties: Comparative Texts, 171
Toman, Michael A., 183
Toxic Substances Control Act, United States, 96
Toxic wastes, 110–11, 185
Tragedy of the Commons, 168
Transboundary problems, 123
Transcript in English of Internet Chat with Commissioner Franz Fischler on the Reform of the Common Fisheries Policy, 180
Treaty of Amsterdam, 30–31, 32, 40, 59, 127, 171
Treaty of Nice, 28, 171
Treaty of Paris, 27
Treaty of Rome, 27, 33, 42, 58, 171
Treaty on European Union, 29–30, 37, 38, 39, 43, 45, 59, 61, 127–35
Tropospheric ozone, 8

UFZ Discussion Papers: European Chemicals Regulation and Its Effect on Innovation, 178, 185
Underdal, Arild, 170
Understanding Climate Change: A Beginner's Guide to the UN Framework Convention and Its Kyoto Protocol, 182
United Nations, 17, 186
United Nations Framework Convention on Climate Change, 80, 100, 103

United States National Environmental Policy Act, 95
Urban development. *See* Decision No. 1141/2001/EC on a Community Framework for cooperation to promote sustainable urban development

Victor, D. G., 170
Vienna Convention on the Law of Treaties, 21
Vig, Norman J., 170
Vogler, John, 169–70

Wallstrom, Margot, 108
Wann, David, 186
Ward, Ian, 171, 174
Waste, 9. *See also* Community Directive on Waste, Directive 75/442/EEC
Waste, Community Directive on, 49
Water Framework Directive, 107–8
Water Policy, European Union Communication on Community, 115
Water Policy, Framework Directive on, 188
Water Pollution, 78
Welsh Brown, Janet, 169–70
What Maastricht Means for Business, 176
White Paper on a Common Transport Policy, 103
Wild Birds Directive Annex I, 73
Wind energy, 82–84
World Commission on Environment and Development, 17–18, 23, 169–70

Young, Oran R., 169–70, 187
Your Voice in Europe, 54